Looking for Trouble:

A Policymaker's Guide to Biosensing

Robert Armstrong, Patricia Coomber, and Stephen Prior

With Ashley Dincher

Center for Technology and National Security Policy

National Defense University

June 2004

The views expressed in this article are those of the authors and do not reflect the official policy or position of the National Defense University, the Department of Defense, or the U.S. Government. All information and sources for this paper were drawn from unclassified materials.

Robert E. Armstrong is a senior research fellow in the Center for Technology and National Security Policy at the National Defense University. Dr. Armstrong may be contacted via e-mail at armstrongre@ndu.edu or by phone at (202) 685-2532.

Patricia K. Coomber is a senior research fellow in the Center for Technology and National Security Policy at the National Defense University. Dr. Coomber serves in the U.S. Air Force as a colonel (select).

Stephen D. Prior, Ph.D., is Director, National Security Health Policy Center at the Potomac Institute for Policy Studies. Dr. Prior also holds a position as a distinguished research fellow in the Center for Technology and National Security Policy at the National Defense Univesity.

Ashley Dincher is an undergraduate student at Bucknell University. In 2003 she served as an intern at the Center for Technology and National Security Policy at the National Defense University.

Executive Summary

Protecting the population against the effects of a bioterrorism attack is one of the most daunting tasks facing government officials. Some of the information required to make informed decisions is highly technical, and even the technical experts do not agree about many of the details or issues involved. This primer is written for the non-technical policymaker and is designed to assist him or her in reaching important decisions regarding how best to help provide early warning of a biological attack.

The authors also present the results of an extensive statistical study that examined the utility of a system-of-systems approach to identifying a bioattack. Using a hypothetical system-of-systems that obtains medically relevant data from 10 sources, the study reaches several conclusions. Among them, that policymakers:

- Reassess efforts currently underway that attempt to capture data from absenteeism reporting, OTC pharmacy sales and medical claims reporting, because their value added may not be worth the cost.
- Increase efforts to collect medical data. These efforts would include, but not be limited to, capturing data from doctors' offices and ER visits, as well as expanded veterinary and agricultural surveillance. Increase data collection from medical website visits and nurse helplines.
- Reassess current plans to significantly increase the number of biosensors deployed as part of both the BioWatch and Guardian programs.

Finally, the authors propose testing an innovative approach to monitoring for the presence of biological pathogens. They recommend that the 23,500-strong workforce of law enforcement officers, firefighters and mail carriers in Washington, D.C., be monitored daily by thermal imagers for increases in body temperature. This workforce is uniformly distributed throughout the city and is both inside and outside of buildings, thus avoiding some of the problems cited with the current use of stationary sensor systems.

Table of Contents

APPENDIX A – THE WARGAME
APPENDIX B – METHODOLOGY & STATISTICAL ANALYSIS

What Are We Looking For?

The spectrum of disease concerns shown in figure 1 ranges from disease states potentially caused by bioterrorism or biological warfare agents to emerging diseases that are naturally occurring—exemplified by the agents responsible for Severe Acute Respiratory Syndrome (SARS) and Avian Flu.

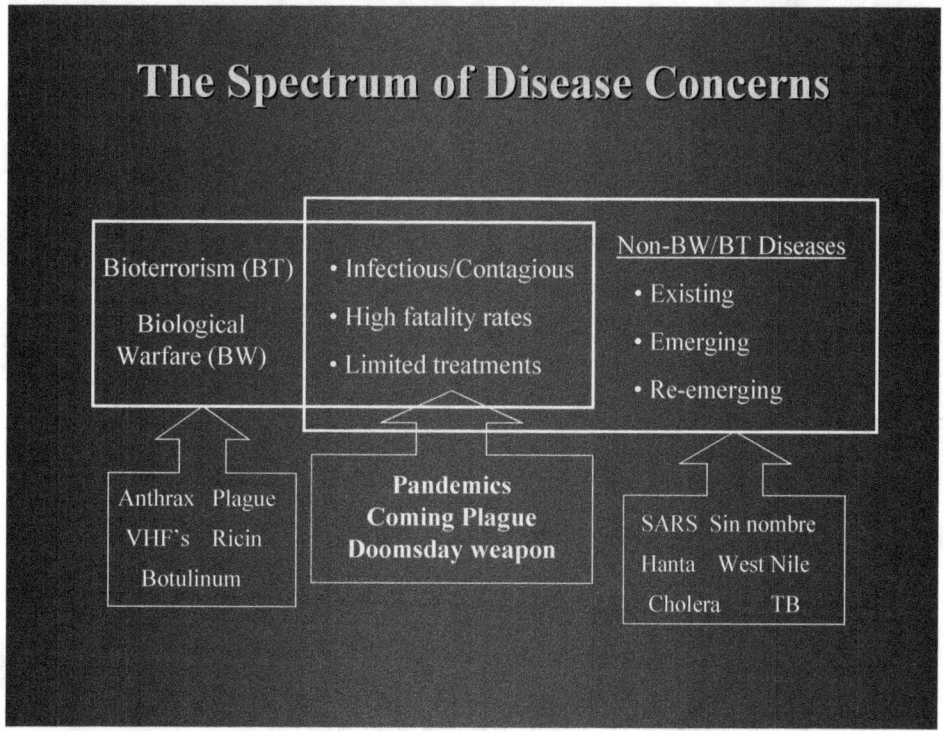

Figure 1 – The spectrum of disease concerns ranges from traditional bioterrorism threats to emerging and re-emerging infectious diseases such as SARS and the West Nile Virus.

Biological agents—usually viruses, bacteria or toxins—are the cause of the majority of the disease states that concern us. A brief survey of basic biology will be helpful in understanding the issues associated with this problem.

Bacteria are living organisms. They contain genetic material and reproduce both sexually (exchanging genetic material) and asexually, (simply splitting in two).[1] Bacteria can be beneficial to humans, as are the billions of bacteria that inhabit our guts and aid in

---

[1] Neil A. Campbell and Jane B. Reece, *Biology.* 6th edition. (San Francisco, CA: Benjamin Cummings, 2002), 340-341.

digestion.[2] Or, they can be quite lethal, regardless of the route of entry or location, as evidenced by the deaths caused by *Bacillus anthracis* during the anthrax attacks.

Bacteria are extremely small. The average bacterium is about one to five micrometers long and about one micrometer wide.[3] (A micrometer is about one one-millionth of a meter. The period at the end of this sentence is about 500 micrometers wide.)

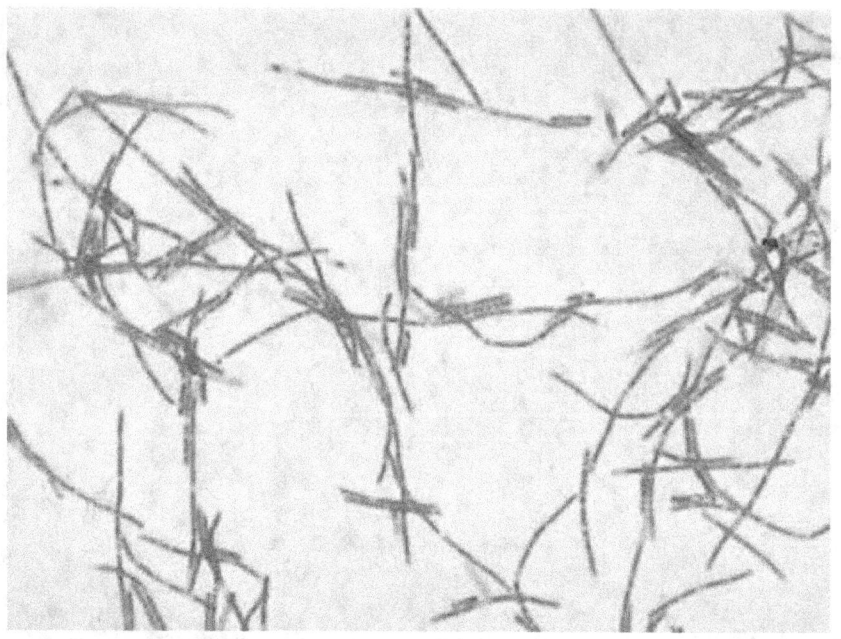

Figure 2 – *Bacillus anthracis*, the bacteria responsible for the infectious disease anthrax. Source: *Public Health Image Library*.

Antibiotics can kill bacteria in a variety of ways. They can attack the cell wall surrounding the bacteria, act on membranes inside the bacteria, disrupt the production of proteins inside the bacteria, or interfere with the manufacture of nucleic acids needed for genetic material inside the bacteria.[4]

Bacteria can cause illness directly, in which case antibiotics may be successful in killing the bacteria before much harm is done to the infected individual.[5] Bacteria also

---

[2] It is interesting to note, however, that the *E. coli* bacteria that inhabit our gut can be lethal if they make their way to other parts of our bodies.
[3] Campbell and Reece, 329.
[4] Ibid., 529.
[5] Ibid., 344-345.

can produce toxins that can cause illness or death.[6] The toxins are essentially by-products from bacterial growth and are not susceptible to the antibiotics. Killing the bacteria before too much toxin is produced is often the only approach to preventing toxic effects, as antitoxins do not exist for many of the bacterial toxins. This is the case for anthrax, for example.

Toxins also can be derived from other living organisms and used directly in an attack. This is actually an example of chemical warfare. Ricin is a toxin derived from the castor plant. No antitoxin exists, and, because a toxin is not a living organism, antibiotics are of no use against it. Many toxins exist and can be aerosolized and used in a fine spray as a weapon. Toxins used as weapons would be of extremely small size—orders of magnitude smaller than a bacterium.

Figure 3 – A computer model of ricin, the toxic protein made from castor beans. *Source: http://www.nbc-med.org.*

---

[6] Campbell and Reece, 541.

Viruses "live" on the edge of being animate and inanimate. They are nothing more than genetic material surrounded by a protective coat of protein.[7] They have no metabolic functions and are incapable of reproducing on their own. However, when they infect a cell, they take over the cell's reproductive machinery and cause it to manufacture new viruses.[8] Viruses can be rather benign, as is the cold virus, or quite deadly, as is the human immunodeficiency virus (HIV). Because viruses are not alive in the sense that bacteria are, antibiotics are of no use against them. Some antiviral drugs exist, but are usually rather specific to selected viruses. Vaccination is the usual way we protect humans against viruses, although vaccines do not exist for all viruses.

Viruses are exceedingly small—much smaller than bacteria. A single virus particle is about 65 nanometers[9]—a nanometer being one one-billionth of a meter.

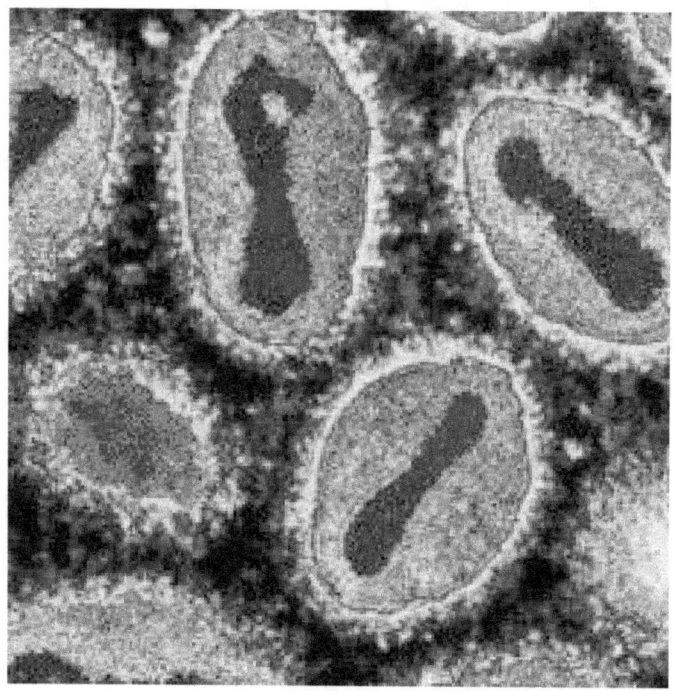

Figure 4 – Smallpox is caused by the variola virus shown above. The virus emerged in human populations thousands of years ago. *Source: http://microbes.historique.net/smallpox.html.*

---

[7] Campbell and Reece, 329.
[8] Ibid., 330.
[9] Ibid., 329.

Other possible agents exist for a biological attack, but in general, bacteria, toxins, and viruses are the most likely candidates. They also are the most likely agents of a naturally occurring biological incident. The amount of toxin or the number of bacteria or number of virus particles needed to make a person ill varies as to the type of toxin/organism being considered and the health of the individual in question. It can be quite small, however. Indeed, many of the biological agents of concern are much more lethal than even our most lethal chemical warfare agents. (See figure 5.)

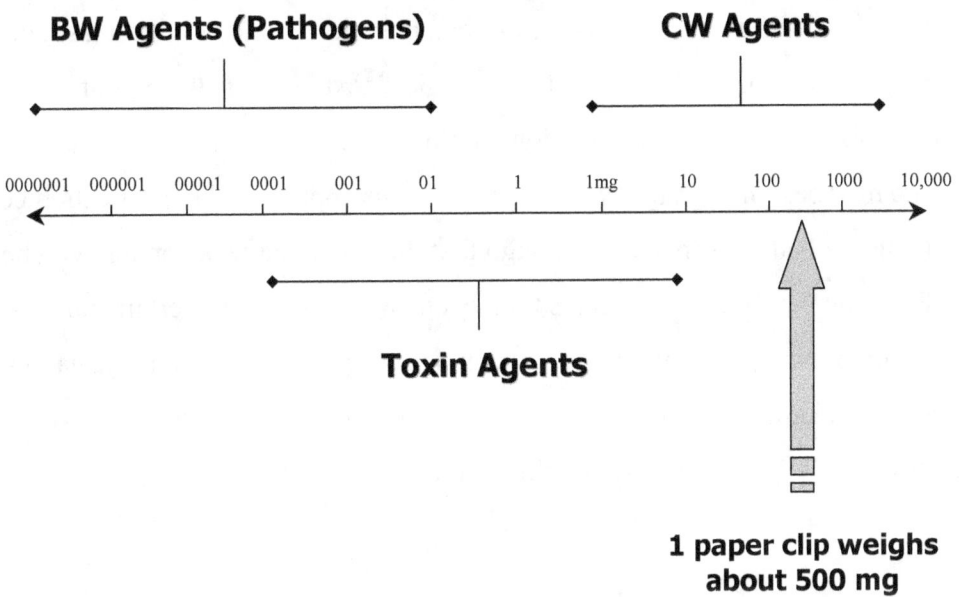

Figure 5 – Comparative toxicity of effective doses of biological agents, toxins and chemical agents.

Also, it is important to note the distinction between something being infectious and something being communicable. A person can become infected with a given virus—say HIV, which we all know to be lethal—but will not be contagious and transmit that virus to other humans by just walking around in the course of daily activities. So HIV is infectious but not contagious, while smallpox is both infectious and contagious.

Thus, when discussing a system for sensing the presence of potentially lethal agents, a system has to be able to detect microorganisms and toxic molecules of extremely small size and possibly in extremely small concentrations.

Finding The Bugs

Ideally, a surveillance system would provide early enough detection and identification that the population could be warned—the *detect-to-warn model*. Receiving adequate warning would allow the population to protect itself by avoiding exposure simply by going indoors or covering one's mouth and nose until in a safe place. Detect-to-warn would require near-real-time detection, however. While this may be viewed as the gold standard, detect-to-warn technologies are largely in the development stage and tend to be quite specific. (For example, most can only detect one or at best a small set of pathogens. See the "Future Biosensor Technologies" text boxes at the end of the chapter for a further discussion of work being done in this field.)

The next best thing is to *detect-to-treat*. In this approach, the notification comes too late to alert all citizens, but soon enough that those who have become ill can be treated. For example, figure 6 indicates that, even with exposure to certain pathogens, there is a window of opportunity within which people can be treated and casualties minimized. This implies the need for a sensing system that can provide positive identification of the pathogen within a timely manner.

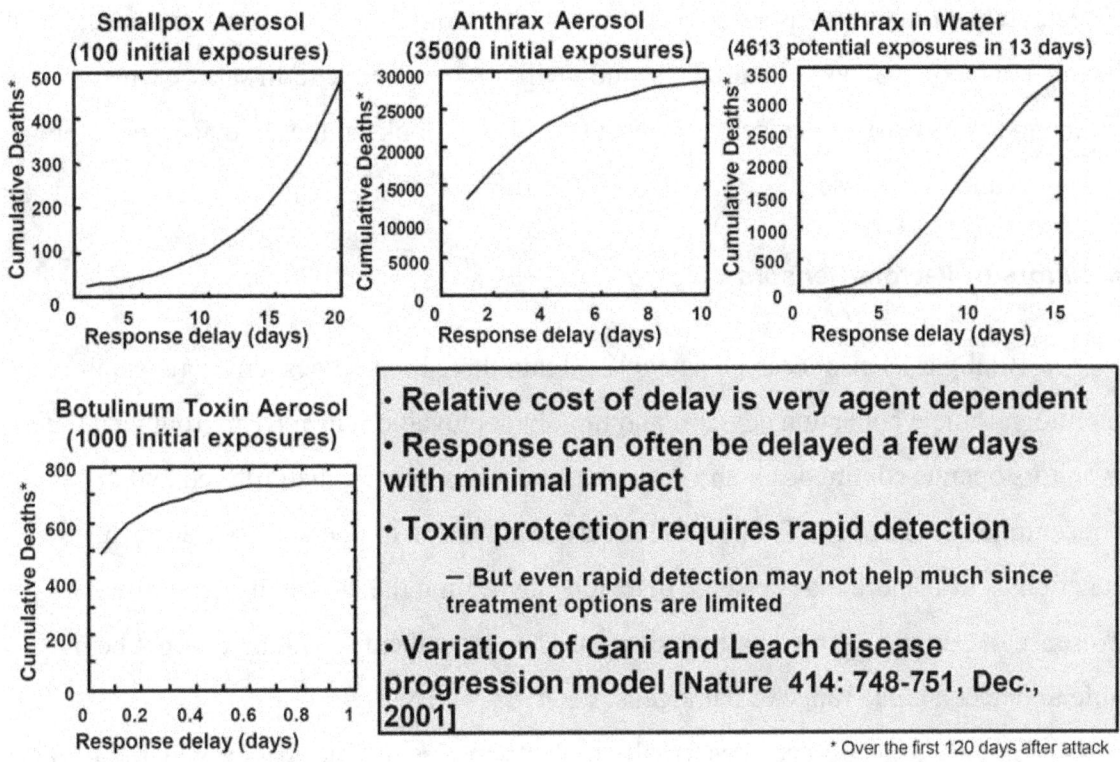

Figure 6 – The impact of response delay on the casualties suffered during a biological attack. Source: Dr. Tim Dasey, MIT Lincoln Laboratory. Received December 19, 2002.

Additionally, a surveillance system should provide sufficient material for independent verification. Such verification is necessary as an additional check on the system, as well as confirmation that the measures taken by authorities are correct and adequate.

A surveillance system should also be able to *detect-to-reassure*. That is, it should continue to monitor during and after an incident to reassure citizens that the levels of the agent in question have subsided enough to discontinue any emergency procedures.

Finally, a system should be able to offer some predictive ability for areas downwind of the initial attack. Not only would this have the effect of providing some detect-to-warn for certain areas, it would provide additional data for officials planning the use and distribution of limited emergency resources.

There are numerous ways to monitor the environment for the presence of biological agents. The most common method currently used is sampling the air with monitors, collectors, or sensors. Generically, all of these systems are sometimes—though inaccurately—referred to as sensors. They are all essentially a form of electro-mechanical device—a box—that takes in air. This study will use the generic term *sensor* when referring to this class of devices. The discussion below looks at these devices and their strengths and weaknesses.

Monitors/Collectors/Sensors

Biological sensors can be categorized into three broad types: environmental monitors, sample collection devices, and rapidly deployable sensors. Environmental monitors operate continuously and draw in air samples that are then filtered and concentrated for analysis. They are relatively inexpensive to operate. However, they function as little more than "change detectors" by testing the environment for the presence of particles of predetermined sizes. They have the disadvantage of not being able to detect a broad range of pathogens.[10]

Sample collection devices usually collect samples on filter paper, which must then be further analyzed. Their chief liability is the cost of operation. Reagents typically cost about $1 per test; some of the reagents have to be maintained under controlled storage conditions, as well. In addition, there is considerable labor involved. While the analysis can be highly specific, the tests are not especially good at identifying novel biological agents.[11]

Rapidly deployable sensors, such as hand-held assays, are limited by the somewhat narrow band of pathogens that they can identify. Their advantage lies in their ability to be dropped on a target area, or easily carried by emergency personnel.[12]

There are common components to all of these devices. First, any sensing system must collect samples from the environment. As discussed above, the size of the items of interest—bacteria, toxins, and viruses—is of such minute scale that, in addition to collecting samples, the samples must be concentrated.

When collecting air samples, it is important to ensure that an adequate amount of air is screened for potential pathogens. Figure 7 uses a range of hypothetical cases to

---

[10] National Institute of Justice, "An Introduction to Biological Agent Detection Equipment for Emergency First Responders," NIJ Guide 101-00, December 2001, 23-25.
[11] Ibid., 23.
[12] Ibid., 26.

8

illustrate this point. For example, if the ID_{50} (Infectious Dose 50, i.e., the dose at which 50 percent of the exposed population will be infected) is 10 organisms, then the sensor will have to collect at least 10 organisms in every 100 liters of air that it samples. The significance of figure 7 is that the level of sensitivity required for a sensor is directly related to the pathogens in question. Thus, one size sampler will not necessarily fit all situations.

Detector Sensitivity Requirements

IF......ID_{50} is 100 organisms

AND..Aerosol retention is 60%

AND..Minute volume is 10 liters

AND..Cloud is on site for 10 min.

ID_{50}	Human	Sensor Reqmt.
10	6 org/100 l	10 org/100 l
100	6 org/10 l	10 org/10 l
1,000	6 org/ l	10 org/ l
10,000	60 org/ l	100 org/ l

100 org. × 60% = 60 org.

10// min. × 10 min. = 100 liters

Must detect 60 org. in 100 liters OR 0.6 org/liter

(or 6 organisms/10 liters)

Figure 7 – Detector sensitivity requirements given that the human must see 10 organisms to retain 6. *Source: Dr. James Valdes, USA SBC Command. Received February 2003.*

Table 1 provides a review of the four most common collection technologies and their respective advantages and disadvantages. The important point to be made from table 1 is that no technology is a clear winner. They are either hampered by the inability to collect large samples, or by the costly requirements for labor or material.

Collection & Sampling Technologies	Advantages	Disadvantages
Cyclone Collectors	• inexpensive & require little maintenance • concentrate contaminants from a large volume of air into a small volume of liquid	• collect all aerosol particles • small, portable devices have problem collecting large enough samples from low concentrations in aerosol
Virtual Impactors	• result in highly concentrated liquid sample • particles of specific size range can be collected	• require a series of probes
Bubbler/Impingers	• can collect very small size particles	• require reservoir of liquid to capture sample
Variable Particle – Size Impactors	• can collect and sort particles of variable size	• require petri dish with growth medium—only suitable for lab environment

Table 1 – Table illustrating the collection and sampling technologies currently available as part of biosensing systems.

Once a sample has been collected, it must then be assessed for the presence or absence of suspicious items. Usually, a "trigger" approach is used. A simple trigger will be activated if there is an increase in the background of particulate matter. If that threshold is reached, then a specific trigger will be activated to investigate the sample for more precise information.[13] The specific trigger may be no more sophisticated than to determine if the suspect material is biological or non-biological.[14]

Table 2 presents five of the most common triggering/detection technologies, with their advantages and disadvantages. As with the collection technologies, no one technology is clearly superior. Note that the disadvantages mostly center on issues involving additional labor, either to perform follow-on tests or to provide maintenance for the equipment in question.

---

[13] Brian M. Sullivan, "Bioterrorism Detection: The Smoke Alarm and the Canary," *Technology Review Journal*, Vol. 11, No. 1, Spring/Summer 2003, pp. 135-141. See <http://www.ms.northropgrumman.com/PDFs/TRJ2003/03ssSullivan.pdf>, accessed March 2004.

[14] Note that in this case, the presence of a toxin would likely go unnoticed.

Triggering & Detecting Technologies	Advantages	Disadvantages
Fluorescence Particle Sizing	• can discriminate between non-biological and biological aerosols	• not good for novel agents, as comparison is made with previously stored calibration curve
Pyrolysis-Gas Chromatography – Ion Mobility Spectrometry (IMS)	• more sensitive than typical mass spectrometry, e.g., allows for detection of bacterial spores	• complicated device involving numerous sensitive pieces (including ionization source)
Flame Photometry & Gas Chromatography (GC)	• can be used for detection of specific compounds • does not usually destroy the sample	• requires careful and frequent calibration • cannot completely characterize a compound
Size & Shape Analysis	• can be used to determine size and general shape of particles • can analyze between 5-10 $\times 10^3$ particles/second	• requires additional fluorescence technology to discriminate between biological and non-biological particles
Flow Cytometry	• fast sample preparation and analysis • simple to operate	• most instruments built for lab environment • extensive maintenance requirements

Table 2 – Table illustrating the triggering and detection technologies currently available as part of biosensing systems.

The final step in any sensing system is the identification of the suspect material. Table 3 lists four of the most common technologies used. The first three—mass spectrometry, antibody based, and DNA-based—all have different variations that are used, but the advantages/disadvantages listed are reasonable summaries that capture the technologies in general. The most significant point of table 3 is the high degree of specificity of these technologies. Thus, sensors using these technologies will be limited to searching for a fairly specific list of possible agents. Novel agents could easily go undetected.

Identification Technologies	Advantages	Disadvantages
Mass Spectrometry	• can provide information on molecular structure	• depending upon which MS technique is used, varying results are obtained
Antibody-based	• highly specific identification of agent	• not able to quickly recognize novel agent
DNA-based	• highly sensitive (very small amounts can be detected) • highly specific (unique DNA/RNA sections)	• difficulty in isolating DNA samples • nucleic acid probes degrade with time
Raman Scattering	• highly specific and accurate for known agents	• not able to identify novel agent

Table 3 – Table illustrating the identification technologies currently available as part of biosensing systems.

In looking across tables 1, 2, and 3, it is clear that no single collection, detection, or identification technology is superior. Nor is any combination of technologies superior. Moreover, the most reliable technologies are also the slowest—keeping them limited to detect-to-treat at best.

On the positive side, the current state of the art can provide systems that will identify the most likely agents and can be used for specific scenarios. Also, in assessing the current technologies it appears that no theoretical barriers restrict future developments. Thus, the gold standard of detect-to-warn eventually may be achievable.

In an ideal world, a biosurveillance system would embody the following characteristics: rapid; sensitive; specific; easy to use; automated; able to use diverse sample types (blood, serum, urine, food, air, water, soil); compact and portable; self-powered; low-cost; rugged. As tables 1, 2, and 3 reveal, this is not an ideal world. But, as stated above, there are no theoretical barriers to eventually achieving a system to meet these criteria.

How Many Boxes Do We Need?

Assume an ideal world and the ability to build a box that meets all of our criteria. Could we afford to build, operate and maintain all that we would need? Before answering that question, let us consider the origin of our ideas on the use of these devices.

Most of our thinking on how to deploy a biosurveillance system for the protection of civilian populations has come from the military use of chemical detectors on the battlefield. The problems are not the same, for several reasons. First, chemical agents act within minutes, if not seconds. Thus, chemical sensors *must* operate as detect-to-warn, if they are to have any utility.[15] Second, chemical weapons often are accompanied by a distinctly visible cloud and/or odor. Those in the downwind area may actually be able to see the approaching danger, even without benefit of sensors.[16] Third, a chemical cloud is effective only in high concentrations and where terrain and meteorological conditions are favorable. Usually, the cloud is not a threat if it is much more than a few meters off the ground.[17] Finally, it is anticipated that the opposing force on the battlefield may use chemical weapons. Typically, the opposing force winds up—at least in part—arrayed in a linear fashion to the front of the friendly forces.[18] There is little question as to the direction from which an attack will come and where to place the sensors.

---

[15] David Ruppe**,** "Rapid, Accurate Biological Attack Detection Capability Is Years Away, Experts Say," *Global Security Newswire*, October 22, 2003. See <http://www.nti.org/d_newswire/issues/2003/10/22/cc3cd030-62a3-46bf-bfe2-ee3e9c49a2c2_html>.

[16] "Fed's Sniffing Devices Effective?" *Associated Press*, Philadelphia, July 17, 2003. See <http://www.cbsnews.com/stories/2003/07/17/tech/main563765.shtml>, accessed March 2004.

[17] Army Online Training Course, FM 3-6. See <http://155.217.58.58/cgi-bin/atdl.dll/fm/3-6/CH1.PDF>, accessed March 2004.

[18] CAPT Sean E. Hynes, USMC, and Neil C. Rowe, "Multi-Agent Simulation for Assessing Massive Sensor Deployment," MOVES Institute, U.S. Navy Postgraduate School. See <http://www.cs_nps.navy.mil/people/faculty/rowe/oldstudents/hynespap_htm#_ftn1>, accessed March 2004.

Figure 8 – On the battlefield, the likely direction of a chemical attack is easily assumed, as opposing forces typically array themselves in linear fashion. *Source: http://www.conmon.com/gallery/albums/2003_04/030403_iraq_tanks.jpg.*

Compare this with biological agents. Biological agents, as seen in figure 9, can take days to produce symptoms.

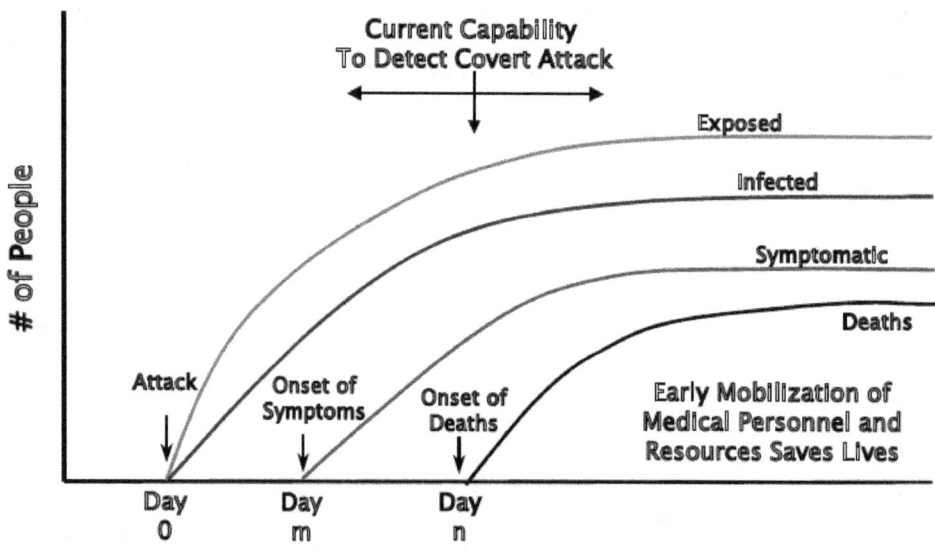

Figure 9 – The characteristics of a generic infectious disease (bioterrorism agent). For example, smallpox generally does not become symptomatic until after an incubation period that averages 12-14 days.

Thus, a surveillance system can function as detect-to-treat and still be useful. Biological agents would be odorless and invisible. Moreover, if the agents were contagious, then a secondary wave could be established by the presence of infected individuals. (For example, on average, in an unprotected population, every person with smallpox will infect another four to seven others.[19]) There need not be high concentrations of agent to create casualties. A biological attack could be effective in any terrain setting. For example, using a contagious agent, an attack could be successful if it were conducted by the simple introduction of infected individuals. Moreover, the geometry of an urban setting is distinctly different from that encountered in a non-urban battlefield. Obstacles abound and microclimate conditions prevail in the canyons formed by tall buildings. (This would argue against the use of line attacks with large clouds of agents.) Finally, the direction of attack by the "opposing force" would be unknown, causing a city to consider ringing itself with a surveillance system—like an army surrounded on all sides.

Figure 10 – The canyons of urban areas create microclimates and geometries distinctly different from traditional battlefields. *Source: http://digilander.libero.it/travelphoto/New%20York/skyscrapers.htm.*

[19] Martin I. Meltzer, et al., "Modeling Potential Responses to Smallpox as a Bioterrorist Weapon," Center for Disease Control and Prevention, Atlanta, GA, December 2001. See <http://www.cdc.gov/ncidod/EID/vol7no6/meltzer_appendix1.htm>, accessed March 2004.

What about the number of boxes required? A recent JASON report concluded that it is not realistic to undertake a nationwide, blanket deployment of biosensors.[20] With current technologies, JASON estimated that approximately one sensor per square kilometer would be required to protect the U.S. population—at an annual cost of $10-15 billion.[21]

In one small study designed to assess the number of sensors needed to protect a moderate-size military base, the numbers suggested that as many as 1,000 sensors might be required to detect an anthrax attack disseminated by a cloud over an area of about 180 square kilometers—roughly the size of Washington, DC.[23]

The real message from this analysis is that one size does not fit all. Moreover, it reconfirms the earlier assessment that there is no biosurveillance technology—nor group of technologies—currently available that provides a clearly superior approach. There is no denying that biosensors are and will continue to be an important part of biosecurity. But, as seen from the brief discussion of the number of sensors required and their costs, a strategy based solely on the use of sensors could quickly become cost prohibitive; especially for a major metropolitan area. The near-term role of biosensors will most likely be as detect-to-treat devices deployed at specific high-value sites.

[20] JASON, "Biodetection Architectures," The MITRE Corporation, McLean, VA, February 2003. See <http://www.fas.org/irp/agency/dod/jason/biodet.pdf>, accessed December 2003. JASON is a group of distinguished defense consultants.
[21] Ibid., 5.
[23] Timothy Dasey, PhD, MIT Lincoln Laboratory, personal communication.

WHAT DOES IT MEAN IF THE ALARM GOES OFF?... OR IF IT DOESN'T GO OFF!

What is the appropriate response if a sensor tells us of the presence of a bioagent? Can we be certain it is a real event, or is it a false alarm? Is it a true positive, or a false positive? And, if the alarm does not go off, is that a true negative and we don't need to worry, or is it a false negative and the alarm should have sounded, but the sensor missed the event?

Significant consequences arise from responding to false positives —loss of public confidence in the system, financial and health costs associated with evacuations, quarantines, and with too many "crying wolf events" potential failures to respond to an actual attack, etc. Similarly, false negatives may lead to significant losses of life and of public confidence.

Detecting the presence of biological agents with sensors can be viewed in a technical sense as a problem in signal detection theory. There is a signal—the presence of certain biological information. There are various sources of "noise" in the background—any number of things that can confuse, mask or distort our ability to detect the signal for which we are searching. For example, in some systems, pollen grains can become "biological noise" that disrupt our ability to detect the signal for which we are searching.

The more sensitive a sensor—i.e., the more discriminating it is with respect to finding the signal amongst the background noise—the greater the probability of false alarms. This relationship is characterized by a type of curve known as a receiver operating characteristics curve (ROC curve). Originally developed for assessing radar, ROC curves have applicability to any number of such tasks. Figure 11 shows a generalized ROC curve.

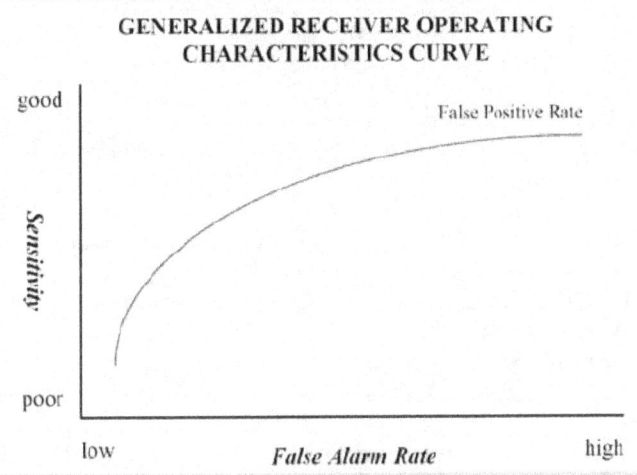

Figure 11 – Generalized receiver operating characteristics (ROC) curve.

Designers face an inherent trade-off when designing sensor systems: the more sensitive they make them, the greater the probability that they will misinterpret the signal-to-noise ratio and give you a false positive—with all of the attendant consequences suggested above. Reducing the false positive rate, however, reduces the ability of the sensor to discriminate the signal from background noise; thus giving you the potential for false negatives.

FUTURE BIOSENSOR TECHNOLOGIES

Cellular Analysis and Notification of Antigen Risks and Yields (CANARY) –
Information taken from Todd H. Rider, et al., "A B Cell-Based Sensor for Rapid Identification of Pathogens," Science 2003, 301: 213-215.

Researchers at MIT's Lincoln Laboratories began work in 1997 on the CANARY (Cellular Analysis and Notification of Antigen Risks and Yields) project. It involves the use of B-lymphocytes, a type of white blood cell that our bodies use against bacterial and viral invaders. These cells are already designed by nature to search for any bacteria and viruses very rapidly. In the laboratory, they are given the ability to glow in the presence of certain contaminants by adding a luminescence gene from jellyfish. The actual detectors are pathogen specific antibodies within the B cells that trigger a burst of calcium when an agent is detected. Within seconds, the calcium activates a bioluminescent protein that causes the whole cell to glow. A device termed a luminometer is used to analyze the light-emitting cell. Within the luminometer the cells are kept alive in test tubes and their response is displayed on a computer readout. The system has already been tested successfully against a list of biological agents, including anthrax, smallpox, plague, tularemia and encephalitis.

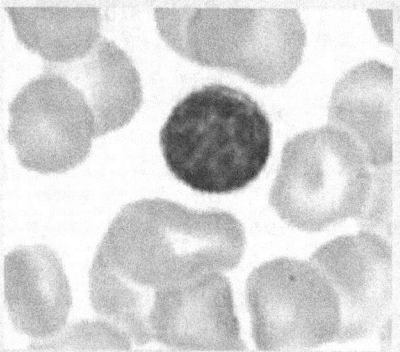

Figure 12 – B cells from the human immune system. *Source:*
http://www3.kmu.ac.jp/anat1/edu/histology/general/blood/bcell.jpeg.

The CANARY bio-agent forensic analysis of body fluids would be useful for monitoring air, water, and contaminated surfaces as well as body fluids. It is expected to detect more rapidly and with greater sensitivity than conventional sensors that are based on chemical reactions. These chemical reactions can take several hours to complete and the sensors can require several thousands of particles for detection. By comparison, CANARY has been able detect as few as 50 colony-forming units of the plague bacterium in less than three minutes. Furthermore, unlike many existing sensors, CANARY would not require advanced training for operation. Consequently, MIT researchers foresee a variety of applications for the system. For instance, in medical diagnostics, it could be used to immediately separate those patients suffering from the symptoms of a cold from those with SARS. In the environment, it could be used to test water and air quality for pathogens both inside and outdoors. In the event of an emergency, suspicious substances on the street, subways, or airports could be tested quickly.

FUTURE BIOSENSOR TECHNOLOGIES

Light Detection And Ranging (LIDAR) –

Information taken from Petter Weibring, et al., "Versatile mobile lidar system for environmental monitoring," Applied Optics 2003, 42: 3583-3594.

Light detection and ranging (LIDAR) is a tool for cloud detection and recognition based on the same physical principles as radar, except instead of bouncing longer wavelength radio waves off a target, higher energy light waves are used. An acronym for "Light Detection And Ranging," LIDAR is occasionally attributed to "Laser Identification and Ranging" by those who want to emphasize the recognition feature. Using lasers that generate light waves in the infrared, the ultraviolet and the visible portion of the electromagnetic spectrum, the multiple energy wavelengths of LIDAR furnish more detailed information, including three-dimensional imaging. Limitations on detection distance and resolution are due to the collection and processing portions of the detector. The more specific the level of data desired, the closer the instruments must be located to the cloud.

Under controlled conditions, detection of aerosolized clouds at long distances has been achieved. The drawbacks are primarily financial and the current limited distance capability. LIDAR instruments are not cheap - costing about $4,000 for a simple LIDAR used for speed monitoring.

The U.S. Army's Long Range Biological Standoff Detection System (LR-BSDS) uses LIDAR-based technology on an unmodified UH-60 Blackhawk helicopter to detect aerosol clouds from long distances. The Short Range Biological Standoff Detection System (SR-BSDS) combines infrared LIDAR with ultraviolet light reflectance (UV). The latter provides enhanced discrimination capabilities. Biological agents can be distinguished from non-biological material based on the excitation of the intracellular fluorescent compounds. The most commonly targeted compounds are the amino acid tryptophan, the coenzyme nicotinamide adenine dinucleotide (NADH), the cellular energy storage molecule adenosine triphosphate (ATP) and the vitamin riboflavin. Identification of these compounds verifies that the sample is biological in origin. Possible false positives include pollen, molds, organic excreta and certain agricultural fertilizers based on decaying organic matter.

Figure 13 – US Army's Long Range Biological Detection System mounted on Blackhawk helicopter. *Source: http://www.lanl.gov/orgs/dod/images/Heli-Lidar.jpeg.*

19

FUTURE BIOSENSOR TECHNOLOGIES

Better, Cheaper, Faster –

Information taken from Olga Kharif, "A Sharper Nose for Danger," Business Week Online, May 25, 2003.

A variety of sensors are beginning to enter the marketplace with the potential to reduce significantly the cost of constant surveillance. The present family of sensors costs about $2 million per city, per year, according to officials at the Department of Homeland Security. Most of the current costs are related to labor associated with collecting the filter papers from air samplers and testing them in a laboratory for the presence of pathogens.

Frances Ligler, a senior scientist at the Naval Research Laboratory, has developed a shoebox-size detector that eventually could screen simultaneously for 12 times as many pathogens as today's devices. Such so-called microarrays operate at the intersection of physics and biology. Using laser light to illuminate the samples, the device can identify specific life forms—bacteria and viruses—and toxic proteins—toxins—by interpreting the resulting fluorescence patterns. Such sensors have several advantages, including lower false-positive/false-negative rates. In addition, bacterial samples are not destroyed in the process, leaving them available for further testing, e.g., to determine whether or not they are resistant to antibiotics.

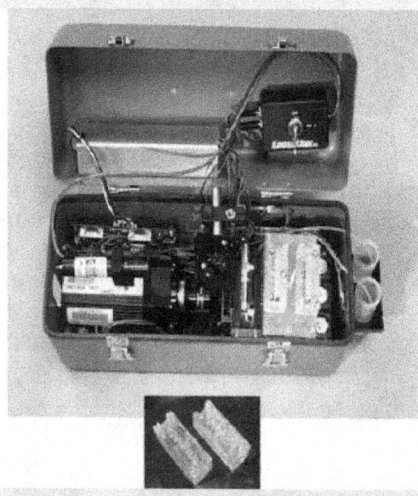

Figure 14 – The most recent version of the NRL array biosensor with integrated optics and fluidics components. Six-reservoir modules for holding samples and fluorescent tracer reagents are shown separately below. *Source: Dr. Frances S. Ligler and Chris R. Taitt, "The Array Biosensors." Received June 1, 2004.*

Several other approaches are also moving towards commercialization. One Boston-area company is developing a reusable tape to which suspect molecules will bind. The molecules can be made to fluoresce, using low-level light. The tape can be rewound and used again.

As biosensor development continues, and their purchase and operating costs decline, their use will expand into other sectors. Food safety and drug manufacture are two areas in particular that will benefit from the ability to rapidly screen for pathogens and toxins.

Other Ways To Find The Bugs

It is not necessary to physically capture and identify a given biological agent in order to determine its presence in a population. Other indicators are available that can provide the initial observation of an agent's presence or provide information that can be used in conjunction with sensor data to confirm an agent. Following is a brief description of data sources in varying stages of development that could be used in an integrated, broad, biosecurity reporting system.

Prodromic Data

Prodromic data is information gained prior to the recognition of disease symptoms in organisms. An example of such surveillance may be the monitoring of immunological markers found in the blood of an organism that does not yet display any symptoms of illness.[24] These markers are some of the earliest indicators, appearing almost instantaneously when the body's defenses are activated. Advances in immunology, molecular biology, and genetics have opened new possibilities for recognizing biological markers to diagnose a range of illnesses, from diabetes[25] to cancer,[26] before symptoms are noticeable.

Much of the work being done in prodromic data collection is still experimental and being carried out in various immunology laboratories around the country. The task of finding specific indicators is a challenge, because the innate human immune response is, for good reason, a rather generic one. (See figure 15 for a simplified view of how the immune system works.) The body is exposed to thousands of harmful triggers called antigens and must be able to respond to each effectively. For our benefit, the immune system cells produce an adaptive immune response that creates long-term, antigen-

---

[24] Farzad Mostashari, Adam Karpati, "Towards a theoretical (and practical) framework for prodromic surveillance." International Conference on Emerging Infectious Diseases, Atlanta. March 24–27, 2002.
[25] Daniel E. Casey, et al., "Optimizing Treatment for Patients with Schizophrenia: Targeting Positive Outcomes," The Center for Health Care Education, 2003. See <http://www.medscape.com/viewprogram/2589_pnt>, accessed February 2004.
[26] University of Maryland Medical System website, "What Tests Indicate the Extent of Existing Prostate Cancer," 2001. See <http://www.umm.edu/patiented/articles/what_tests_indicate_extent_of_existing_prostate_cancer_000033 8_htm>, accessed February 2004.

specific proteins called antibodies.[27] These basic components of an immune response are very likely not the only response the body produces when exposed to disease agents. Researchers at the National Institute of Allergies and Infectious Diseases (NIAID) believe that a number of potential markers exist and have begun organized efforts to effectively evaluate their utility.[28]

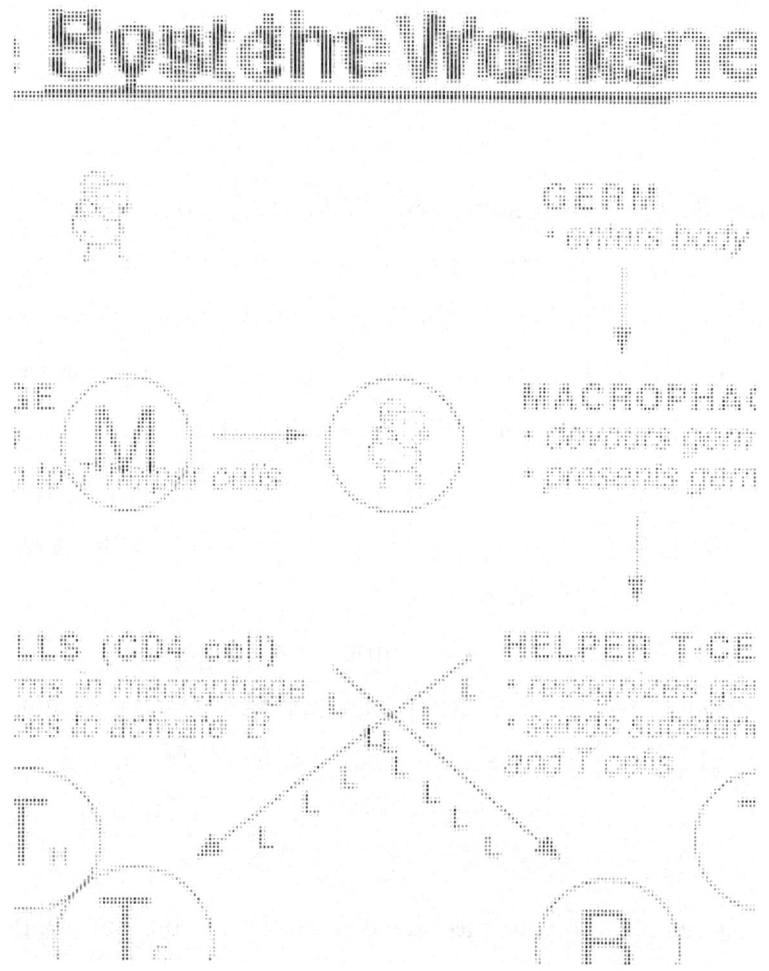

Figure 15 – Schematic view of human immune system. Some researchers suggest monitoring early immune response as a way of detecting bioattacks. *Source: NYC Comprehensive Health Curriculum.*

---

[27] Neil A. Campbell and Jane B. Reece, *Biology*. 6th edition. (San Francisco, CA: Benjamin Cummings, 2002), 904-905.
[28] "CombiMatrix Announces Commercial Launch of CustomArray, a Fully Customizable DNA Array Platform," Business Wire, *BioExchange*, November 6, 2003. See <http://www.bioexchange.com/news/news_page.cfm?id=18816>, accessed March 2004.

Another intriguing application of prodromic surveillance involves the development of a rapid, sensitive, and reliable method to diagnose respiratory infections to rapidly distinguish between infected and non-infected persons.[29] Researchers are developing new, rapid techniques to analyze proteins in the breath of persons with respiratory infections, such as inhalation anthrax. The new techniques take seconds to minutes to accomplish, compared to current laboratory assays that take hours to perform.[30] The current research focus in this area is to define the proteins in exhaled air and in body fluids, such as saliva and urine. Researchers also are examining the immune responses that occur in the lung in response to exposure to biological agents. Rapid diagnosis using breath analysis would be a vital capability during a biological warfare attack, because hospital emergency rooms could be overwhelmed by the "worried well." Also complicating diagnosis is the fact that initial symptoms for a biological agent infection are generally indistinguishable from symptoms for a common cold or flu. Early and immediate treatment of infected persons would improve their prognosis for recovery.

Syndromic Surveillance

Syndromic surveillance refers to the observation of signs or symptoms that characterize an abnormality, such as a disease outbreak in a population. This is a broad designation of the term meant to include observations by medical professionals and the public health sector, as well as from pharmaceutical sales, medical claims reporting, and veterinary surveillance. All can provide valuable information for biological detection. The goal of syndromic surveillance is essentially to shorten the delay between the appearance of the first cases and intervention.[31] This would minimize person-to-person transmission and provide for timely treatment. The value of syndromic surveillance was evident in the postal anthrax attacks in 2001. During the incident, the patients who survived were those who were diagnosed and treated early.[32] In fact, six of eleven

---

[29] http://www.darpa.mil/dso/thrust/biosci/ADVDIAG/Programs/scranton_jhu.html
[30] Johns Hopkins University Applied Physics Laboratory homepage
http://www.jhuapl.edu/programs/rtdc/Pathogens/RapidDetectionAndId html
[31] Stacy Hall, seminar remarks at the *National Syndromic Surveillance Conference* held by the New York Academy of Medicine, September 24 2002, access at
<http://www.nyam.org/events/syndromicconference/2002/presentationpdf/stacy_hall.pdf>.
[32] Ibid., 10.

patients with acute inhalation anthrax survived because they were diagnosed early and received immediate treatment with antibiotics. The early diagnosis and quick response of public health professionals prevented a much greater tragedy.

Health care professionals gather and record valuable information that could be crucial in detecting an attack, if it were quickly monitored and interpreted. Hospital facilities use computer systems to compile data from departments throughout their organization. This includes registration and billing information, clinical observations, emergency room management, radiology, laboratory, and pharmacy services. Since the amount of information recorded is significantly large, computer programs are typically used to maintain organization and accessibility for personnel. Electronic medical records systems are an example of such programs. If these programs were easily accessible for public health monitoring, they would be valuable tools for disease and bioterrorism surveillance.

Unfortunately, comprehensive medical records systems are not widely implemented in hospitals nationwide and are even less frequently used in physicians' offices.[33] In hospitals that do use computer databases, the information is rarely linked or shared with other hospital record systems.

Local and state health departments gather information directly through their own investigations as well as from healthcare professionals. In fact, epidemiologists have been systematically collecting data about disease outbreaks for decades. Currently, the Federal Government uses a few systems for recording a core set of data. Fifty states, the District of Columbia, New York City, and five U.S. territories all gather information via the National Electronic Telecommunications Surveillance System (NETSS) and transmit it to the Centers for Disease Control.[34] There are other disease surveillance systems that collect reports on single diseases, such as the National Malaria Surveillance System. Unfortunately, while the content of the information has great potential, and is a valuable aspect of confirmation, the overall capability of these systems for real-time biological detection is rather low.

---

[33] Hall, 3.
[34] Ibid., 3.

Both medical and public health surveillance are considered to be rather slow indicators of a biological attack. It was mentioned above that a CDC study showed these sectors as being a promising source of biological detection. While the proportion of outbreaks reported was high, the length of time between the first reported case and the identification of the problem sometimes exceeded a week or even two weeks.[35] Much of this time lag is a result of the poor system of communication between healthcare enterprises and public health monitors. There are few direct links between hospital clinical information systems and the public health sector. The current method of data transfer is usually as follows: frontline healthcare professionals identify a reportable case, for which they fill out paper-based data collection forms that are sent to the local health department. From the local department, the data are either copied and filed to the state or transmitted through a computerized electronic data management system. Finally, at the state level, the data are manually entered into an electronic system.[36] Thus, the data become electronically available days to weeks later in a rather unorganized format, because there are no standard methods for public health reporting.

In December 2003, the Institute of Medicine (part of the National Academy of Sciences) called for hospitals and physicians to adopt electronic record-keeping systems that could form the basis for a nationwide network of patient information.[37] The Institute stated that the government would set the standards but would not direct what software clinics and hospitals should buy. Although requirements for participation in a national information network would not be mandatory, involvement would eventually become a prerequisite for participating in such programs as Medicare. A nationwide network would allow constant disease surveillance, which would be valuable in detecting and responding to bioterrorism incidents.

New public health surveillance systems are being designed to minimize this lag by improving and integrating electronic disease surveillance systems. One project now being implemented by the CDC is the National Electronic Disease Surveillance System (NEDSS). This network is a collaborative effort between CDC and state health

---

[35] Hall, 12.

[36] Centers for Disease Control and Prevention. "Supporting Public Health Surveillance," accessed at <http://www.cdc.gov/nedss/>.

[37] Doctors Advised to Keep Records Electronically" BizReport.com; November 21, 2003; <http://www.bizreport.com/article.php?art_id=5601>.

departments to create an integrated and standardized electronic information system for communicable diseases.[38] The system would attempt to ease the burden on health care providers with standardized data collection and would enable earlier recognition of a problem through automatic electronic case reports to the state level.[39]

Another system that is in compliance with NEDSS is Real-time Outbreak Disease Surveillance (RODS). The RODS project is a collaboration of health departments, hospitals and medical centers, foundations, and industries throughout Pennsylvania and northwestern Utah.[40] It is currently in operation and receives real-time data from emergency departments in these regions to be examined for incidents that are suggestive of disease outbreaks.

Computer surveillance systems such as these can report symptoms from emergency rooms or clinics that characterize a possible biological attack. However, it is difficult to choose which symptoms should cause alarm, since many illnesses show similar signs. For instance, anthrax cases often report a fever, chest pains, fatigue, cough, and an abnormal chest x-ray. Yet, the same symptoms are often recorded in flu cases and many viral and bacterial respiratory infections. If these symptoms were allowed to signal possible anthrax cases, confirmatory labs would be overwhelmed.

Complementary to the speed of reporting is accuracy of the information. The certainty of any indicator is vital, since false positive or negative results can be extremely costly both to human health and the economy. When using healthcare data for syndromic surveillance, one may be able to lessen the occurrence of false positives by considering a threshold level for detection that accounts for annual events, such as flu and allergy season. On the other hand, an outbreak such as SARS may initially go almost completely unnoticed if the rise in flu-like symptoms coincides with the number of expected flu cases during flu-season.[41]

A similar problem arises if the spread of the outbreak occurs slowly. A simulation done by RAND Statistics Group showed very positive results for early detection in the

---

[38] *Patient Safety: Achieving a New Standard for Care*; Philip Aspden, Janet M. Corrigan, Julie Wolcott, Shari M. Erickson, *Editors*, Committee on Data Standards for Patient Safety; Nov 20, 2003, 18.
[39] Ibid.,18.
[40] RODS Homepage, access at <http://www.health.pitt.edu/rods/>
[41] Mike Stoto, RAND, seminar remarks at the *Conference on Statistical Issues in Counterterrorism* held at the Keck Center of the National Academies, 29 May 2003.

event of a "fast outbreak," that is, 18 reported cases in three days. However, when the same number of cases was reported over 9 days, the probability of early detection dropped significantly.[42]

The location of the original reporting can provide another confounding variable for syndromic data from the healthcare field. False positives could appear in regions where certain diseases are endemic. For instance, higher incidence rates for cholera and plague were noted in the western United States and for tularemia in the central United States.[43] Research done by the CDC describes disease-specific trends in demographic characteristics as well as geographic and seasonal distribution of conditions caused by certain biologic agents. Studies such as that done by the CDC, which identify patterns of endemic disease associated with these agents, establish a baseline against which future disease incidence can be compared.[44] As a result, identifying the incidence of unusual diseases would become an easier and more reliable method of surveillance.

In summary, medical and public health surveillance systems contain a large array of information of potential value for early biological detection through syndromic surveillance. Most of this information is simply not collected, transferred, or made available through a means that would create the timely recognition of a problem. New programs, such as NEDSS, may be the answer to this problem. The improvements made for developing an effective syndromic surveillance system would create a stronger public health system as well.

Pharmaceutical Sales

The pharmaceutical sales industry is another possible informant for syndromic surveillance. Existing data bases used for market surveillance may be of use for biological detection. One such system is used for the management of prescription drug benefits plans. Pharmacies use online systems to verify health-care coverage for prescription plans and to assign the prescription to the appropriate supplier.[45] The

---

[42] Stoto, 18.

[43] Chang M, Glynn MK, Groseclose SL, "Endemic, notifiable bioterrorism-related diseases, United States, 1992-1999," *Emerging Infectious Diseases*, 2003 May, accessed at <http://www.cdc.gov/ncidod/EID/vol9no5/02-0477.htm>.

[44] Ibid., 25.

[45] Ibid., 3.

information recorded includes the name and dosage of the drug, age and geographic information of the patient, and, in some cases, a code for diagnosis. The data enters a transactional system usually on a weekly basis and is uploaded to a data warehouse that can be made easily accessible to public health authorities. The system has the potential to define trends in prescription drug activity that could indicate a threat to public health.[46] However, cost and privacy issues could make access to these data difficult.

In addition to prescription drug sales, there may be value in monitoring over-the-counter (OTC) drug sales in detection of outbreaks or possible terrorist attacks. The data may be important as an indirect indicator of illness, if it can signal an outbreak before emergency rooms, health clinics, or even prescription drug surveillance. In theory, this could occur if persons with symptoms of illness buy OTC medications first, rather than rush to the ER or consult their physicians. Records of these sales could then be pooled and a detection algorithm used to identify irregular patterns of purchases.[47] For instance, a surge in the sales of a decongestant or cough medication could indicate an increase in bacterial infections or diseases with flu-like symptoms.

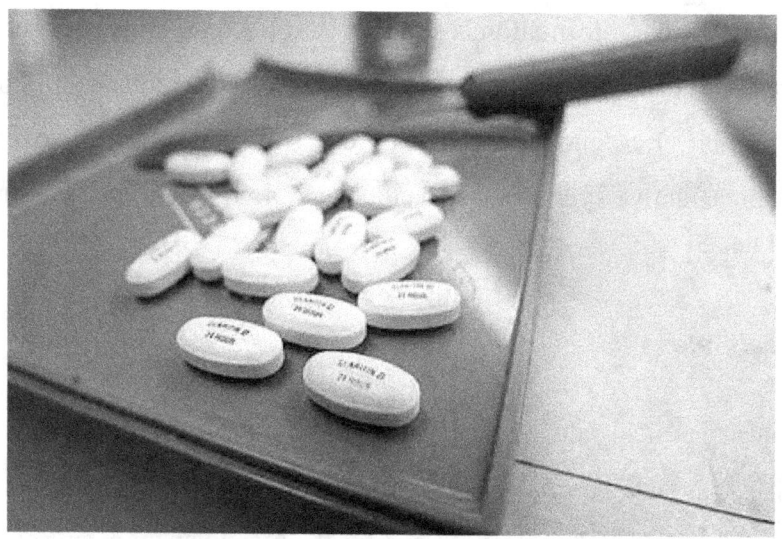

Figure 16 – Over-the-counter (OTC) sales can be monitored as one indicator of a possible bioattack. *Source:* **Honolulu Advertiser,** *March 9, 2002.*

---

[46] Chang, et al., 3.
[47] Goldenberg A, Shmueli G, Caruana R, "Using grocery store data for the detection of bioterrorist attacks", CALD technical report, Carnegie Mellon University, Oct. 2001., access at <http://www-2.cs.cmu.edu/~anya/papers/biodetectionStatMed.pdf>.

Although OTC medication tracking will not facilitate real-time biodetection, monitoring OTC sales may be a piece of evidence for detection. Unlike medical surveillance, OTC sales are not symptom specific. For instance, an attack from a bacterial agent such as anthrax or plague could not be inferred directly from an increase in sales of items, such as Tylenol or Advil, each of which could be used for a variety of symptoms. Perhaps a general sense of the type of illness occurring could be gathered by inferring from a variety of drug sales, such as a combination of cough medicine with decongestant. However, there is little background data indicating the ways in which epidemics manifest themselves in OTC data. Thus, patterns and correlations of these sales and their usefulness for disease surveillance remain uncertain.

One French study did analyze a number of surveillance mechanisms for Influenza A, including OTC sales. They found OTC sales did indeed show a rise during epidemic weeks, but in comparison to other methods, such as emergency rooms visits, the data were less immediate and accurate.[48] Explanations for the delayed signal include the possibility that the majority of the public keeps enough OTC drugs in their homes that they do not need to buy more when symptoms occur. As a result, any rise in sales may indicate these persons are restocking rather than purchasing OTC drugs for their initial symptoms. Furthermore, there are a number of issues that may prove the data to be unreliable or difficult to use. For instance, initially the data recorded contains complete details of the sale, including the exact date and time, customer information, detailed information on each product, prices, etc. Not only is this dataset extremely large, but stores may be unwilling to release such information.[49] Also, sales of certain drugs vary a great deal seasonally, as well as according to pricing policies or on holidays. All of these factors could lead to inaccurate conclusions from the data.

Medical Claims Data Surveillance

Another source for syndromic surveillance data could be medical claims and billing information for detecting public health threats. Claims data are very

---

[48] Philip Quenel, William Dab, Claude Hannoun, and Jean Marie Cohen, "Sensitivity, Specificity, and Predictive Values of Health Service Based indicators for the Surveillance of Influenza A Epidemics," *International Journal of Epidemiology*, (1994) Vol. 23 No. 4 pg. 849-855.

[49] Ibid., 29.

comprehensive. They include demographic information, such as name, age, gender, address, ethnicity, and dependents as well as the patients' complaint, physicians' notes, and tests from laboratories and other medical procedures. The thoroughness of this resource makes it a possible tool for monitoring, detection, and aftermath analysis. However, the submission of information is rarely timely enough for early detection of a biological attack. Medical claims are submitted both on paper and electronically when an insured patient seeks medical attention. They initially enter a transactional system and accumulate until they are sent to a central data warehouse. It is from this warehouse that public health officials could access sufficient claims data that may indicate a possible outbreak. Unfortunately, as 30 days pass between the time the claim is filed and when the information becomes available from the warehouse.[50]

Greater access to claims data could be improved through web-based and other electronic filing procedures for healthcare providers. These features are already becoming a ubiquitous feature of hospital and physician offices in the United States. If this source were pursued, however, the issue of client privacy would have to be addressed. This would likely result in a fee for access to the warehouse and an agreement relieving the warehouse operator of any liability. The cost of providing access to one data warehouse could reach millions of dollars and become even more expensive if used for day-to-day monitoring.[51]

Sentinel Organisms – Animals as Collectors

In the search for the best method to warn of a biological attack, many scientists believe that nature may hold the answer. They argue that biological organisms may actually be more convenient and accurate monitors than machines.[52] Insects could pick up trace amounts of agents in the air or on the ground, acting as collectors. Insects performing this function are referred to as Key Insect Carrier Species (KICS).[53] The

---

[50] Quenel, et al., 3.

[51] James Gallo, "Operations and Maintenance in a Data Warehouse Environment," *DM Review Magazine*, December 2002. See <http://www.dmreview.com/article_sub.cfm?articleId=6118>, accessed March 2004.

[52] Mimi Hall, "Bugs, weeds, houseplants could join the war on terror," USA Today, May 27, 2003. See <http://www.usatoday.com/news/nation/2003-05-27-bugs-cover_x.htm>, accessed March 2004.

[53] See <http://www.talkabouthealthnetwork.com/group/alt.psychology.jung/messages/37768 html>, accessed March 2004.

insects, which appear to be unaffected by the pathogens, are, in turn, collected with devices such as sticky papers and black lights and taken to a lab and tested for hazardous agents. There are at least two significant uses for KICS. First, they could serve for general monitoring of harmful biological agents. Second, they could have a specific detection application if released into a suspected BW storage or production site and recaptured to be examined. One researcher refers to insects used in that fashion as "flying crawling, Q-tips." The advantages of using KICS are that they are a readily available and inexpensive source for collection and monitoring. Having the insects collect bioagents and then collecting them is only part of the process, though. Laboratory analysis of the insects for biological residues, just as with mechanical biocollectors, can be labor intensive and expensive.

Figure 17 – Key Insect Carrier Species—KICS—are proposed as a way to sample for the presence of biological agents. *Source: http://www.rdc.ab.ca/rdc/organic_chemistry/biology/don_wales/insects/images/cockroach.jpg.*

Sentinel Organisms – Plants and Animals as Collectors

Sentinel plants and animals are another alternative to 'black box' detection. Scientists at Penn State University and Colorado State University have been sponsored by the Defense Advanced Research Project Agency (DARPA) to genetically engineer plants that quickly change color if they come into contact with biological or chemical agents.[54]

---

[54] Lakshmi Sandhana, "Greenery on Alert," *The Christian Science Monitor*, May 1, 2003. See <http://search.csmonitor.com/2003/0501/p12s01-stct.html>, accessed January 2004.

Because plants are stationary organisms, they have evolved to be highly sensitive to their environments. Even the simple flowering plant used in their research, *Arabidopsis*, a small weed in the mustard family, is estimated to have nearly 600 receptors and consequent response pathways.[55] Through advanced bioengineering, these plants could be designed to perceive a biological or chemical input and respond in a predetermined way, such as producing a fluorescent protein causing them to glow.

Arabidopsis is a convenient subject for such research because of its simplicity and because its entire genomic sequence has been decoded and is publicly available. If successful, the technology could be applied to plants commonly found in public places, such as shopping malls, offices, roadway medians, a pond, or even in one's home. The plants could be a rapidly responsive and unobtrusive mechanism for detecting the presence of biological and chemical agents.

There are numerous valuable benefits of this research beyond biological detection. If successful, it may also be used for naturally occurring disease and toxin surveillance or pollution monitoring. One aspect of concern, and an issue still not addressed, is the necessity for recharge capability, i.e., the plants would have to be designed to recover quickly to detect subsequent attacks.

The number of smell receptors in a human's nose ranges from 5 million to 15 million, whereas in a dog, it can range from 125 million to 250 million.[56] In addition to more smell receptors, the olfactory portion of a dog's brain is four times larger than a human's.[57] The military is capitalizing on dogs' abilities and training them to detect biological agents. In addition, it has also been noted that patients with smallpox and other biological agent diseases have a specific odor that could, theoretically, be detected by canines.[58]

A very recent application of sentinel animals for biological agent detection occurred during the Iraq War of early 2003. Although the U.S. Marines brought state-of-the-art equipment to warn of a possible chemical or biological attack, their first hint of

---

[55] Kenneth Chang, "Ideal Sensors for Terror Attack Don't Exist Yet," *New York Times*, April 1, 2003. See <http://www.ph.ucla.edu/epi/bioter/idealsensorsnotexist.html>, accessed January 2004.

[56] Maryann Mott, "Dogs of War: Inside the U.S. Military's Canine Corps," for National Geographic News, April 9, 2003, access at
<http://news.nationalgeographic.com/news/2003/04/0409_030409_militarydogs.html>.

[57] Ibid.

[58] See <http://www.cnn.com/2001/COMMUNITY/09/20/siegrist/>, accessed February 2004.

danger may have come from a pigeon. Because pigeons are more sensitive than humans to some biological and chemical agents, dozens of birds were distributed to Marine regiments in Kuwait to warn the troops of a chemical or biological release.[59] Just as canaries once warned miners of the threat of explosive gas, the U.S. military thinks pigeons may once again prove to be the difference between life and death. One sergeant commented that, "I got sensors that cost $12,000 and birds that cost $60 each and I place just as much trust in the bird as the sensor. Anything mechanical can fail or give us wrong readings."[60] Pigeons are good sentinels, because a lethal dose for them will not kill humans. The birds are also capable of detecting toxic gases if an industrial facility or water treatment plant were attacked.[61]

Figure 18 – During the invasion of Iraq, some military units kept pigeons as sentinel animals to warn of the presence of toxic agents. *Source: Associated Press Photo.*

Another approach to using animals as an early warning and detection system is being explored by researchers at the University of Wisconsin-Milwaukee (UWM). The UWM Center for Water Security has received a one million dollar federal grant from the Department of Defense to create fish that glow when exposed to toxic agents. The Center's goal is to help water utilities react to possible threats from biological or

[59] "US Marines Enlist Pigeons to Detect Iraqi Gas," Reuters Health, Camp Inchon, Kuwait, March 14, 2003. See <http://yourhealth.healtheast.org/HealthNews/reuters/NewsStory0314200324 htm>, accessed January 2004.
[60] Ibid.
[61] Ibid.

chemical pathogens. Scientists at the UWM institute are inserting a gene from the firefly into zebrafish. By linking the firefly gene to the zebrafish's DNA, the fish emits a glow when toxic chemicals are present.[62]

Cellular Sentinels

Another approach to using organisms as detectors is at the cellular level, using animal cells or single-celled organisms as indicators for the presence of a chemical or biological agent. The majority of this research has focused on ways to get cells to luminesce when exposed to a toxic agent.

Massachusetts Institute of Technology researchers have engineered mouse immune cells (white blood cells) to glow in response to biological agents.[63] These special cells contain a jellyfish gene for a luminescent protein, as well as antibodies that respond to bacteria and viruses. When the antibodies on the sensor cells detect a pathogen, such as plague or anthrax, a surge of calcium is released. The calcium immediately activates the bioluminescent protein, causing the cell to glow.[64] The cell system, developed with funding from the Defense Advanced Research Projects Agency (DARPA), has been successfully tested against many of the potential bioterrorist diseases such as anthrax, smallpox, plague, tularemia, and encephalitis.[65]

Using this glowing cell system, first responders could test suspicious substances in subways, mailrooms, and airports. The cells would be a valuable tool, because they respond in 30 seconds to 3 minutes, compared to 30 minutes to several hours with current laboratory analysis.[66] Another valuable application of this system is as a diagnostic tool for doctors. Physicians could immediately identify patients with a disease of concern, such as Severe Acute Respiratory Syndrome (SARS), or those just infected with a cold.[67]

---

[62] "Keeping the water safe"; MSNBC/The Business Journal of Milwaukee October 20, 2003, Becca Mader; <http://famulus msnbc.com/famuluscom/bizjournal10-25-010753.asp?bizj=MIL>.

[63] MIT News, "MIT sensor detects pathogens quickly and accurately," July 10, 2003. See <http://web.mit.edu/newsoffice/nr/2003/sensors html>, accessed May 2004.

[64] Todd H. Rider, et al., "A B Cell-Based Sensor for Rapid Identification of Pathogens," *Science* 2003 301: 213-215.

[65] http://www.cnn.com/2003/HEALTH/07/10/finding.germs.ap/

[66] Ibid., 2.

[67] Ibid., 2.

A similar technique is being investigated at Virginia Polytechnic Institute and State University, where researchers are developing a lab-on-a-chip sensor that will detect trace amounts of toxins in water.[68] The sensor is composed of microscopic channels of fluid with *E. coli* bacteria, which release potassium when they come in contact with certain toxic chemicals. The potassium triggers a downstream photosensor to fluoresce as an indicator for a poison (biological or chemical) in the water. Such a sensor would be especially valuable as an early warning system to signal the requirement for additional analysis.[69]

Another approach to water biodetection is being done by scientists at Oak Ridge National Laboratory. The research plays on natural fluorescence in algae living in lakes, rivers, and reservoirs that are primary sources for drinking water.[70] One key advantage in using these algae as biosensors is that they are naturally present in the water. This approach measures characteristics of fluorescence during photosynthesis, which changes when the algae is exposed to toxins in the water. The fluorescence induction data can be used as a real-time tool to detect toxic agents, making this system a significant method for around-the-clock monitoring, early warning, and field-deployable analysis.[71]

Veterinary Surveillance

It is important to recognize that human health may not be the only target for a biological attack. Attacks on animal populations as well as agriculture are not only feasible, but have occurred. In 1915 and 1916 in Maryland, Virginia, and New York, horses and mules were the targets of biological warfare using Glanders and anthrax manufactured in Germany.[72] Additionally, there is compelling evidence that Rhodesian security forces infected cattle with anthrax in 1978-1980, during the Zimbabwe War for Independence, and that the Soviet military used Glanders against horses in Afghanistan in 1982 and 1984.[73] Whether targeted directly or indirectly, animal health is an important

---

[68] Christopher Helman, "Sharpening Our Senses," *Forbes* (April 14, 2003) vol. 171, no. 8, p. 56.
[69] Ibid., p 56.
[70] Oak Ridge National Laboratory FactSheet, March 17, 2003, UT-Battelle.
[71] "Oak Ridge Technology Stands Guard Over Water Supplies, "Inside Energy, June 24, 2002, p17.
[72] Corrie Brown, Mark Thurmond. "Bio- and Agroterror: The Role of the Veterinary Academy." *Journal of Veterinary Medical Education.* Vol 29 no. 1 2002. pp. 1-4.
[73] Ibid., 32.

potential indicator of a biological attack. Consequently, another interest is the role of veterinarians in the detection of such events.

The veterinary profession is founded on service to society through both the protection of animal health and the promotion of human well-being. Their knowledge makes them well suited for the detection of biological weapons, because most recognized bioterrorist agents are zoonotic—that is, they cause diseases in animals that can be transmitted to humans. Such agents include plague, tularemia, anthrax, botulism, and hemorrhagic fevers, the majority of which are already familiar to veterinarians. Many biological pathogens produce similar symptoms in animals and humans. Because veterinarians are uniquely trained to observe such illnesses, most state governments maintain a position in public health specifically for a veterinarian.[74] Not only are veterinarians themselves familiar with diagnosing agents such as anthrax and plague, but veterinary labs also are well-versed in the diagnostics tests for these diseases and would likely prove useful in the event of an outbreak.

Currently, the ability of pets to act as sentinels is the focus of a number of veterinarians at Purdue University, where they are working on software for a national pet health surveillance database. These researchers have recognized that veterinarians may be the first to find symptoms of an outbreak, particularly if an animal population density is high in the affected area or if pets, due to their small size, are more susceptible than humans to agents that may have been released.

Additionally, information that might indicate an outbreak may be collected more promptly by veterinary surveillance than through public health surveillance. This is because of the existence of a centralized, standardized, veterinary hospital network system known as Banfield Veterinary Hospitals. The chain has 310 hospitals in 43 states, and Banfield's vets see an estimated 1 to 2 percent of the nation's cats and dogs—about 2.5 million a year.[75] The chain's hospitals and clinics use the same computer programs and data systems, in which data are entered weekly to a central computer. Such a system is far more effective for identifying a possible anomaly than the unlinked arrangement of physicians and hospitals used for human health surveillance. Veterinarians monitoring

---

[74] Brown and Thurmond, 32.

[75] Elizabeth Weise, "Pet ailments could signal toxic attack," USA Today, 4 April 2003, access at <http://www.delawareonline.com/newsjournal/life/mindbody/2003/04/07petailmentscoul.html>

captive animals have already served as early indicators of an infectious outbreak: a veterinary pathologist at the Bronx Zoo gave the first alarm for the West Nile virus outbreak in the fall of 1999.[76]

Figure 19 – Cats and dogs can be the earliest indicators of a biological attack, and veterinarians will need to play an active role in identifying and communicating potential threats . *Source: http://www.hc-sc.gc.ca/vetdrugs-medsvet/approval_e.html.*

---

[76] Weise, 35.

Wargame

As discussed above, several data streams with biological and medical information are available to policymakers. Individually, these data streams provide varying contributions to addressing the issues of specificity, sensitivity and timeliness. In general, a review of the scientific literature suggests the analytical work done to evaluate their true utility to policymakers has been lacking. One recent research paper concluded: "Most evaluations of detection systems and some evaluations of diagnostic systems for bioterrorism responses are critically deficient."[77] Indeed, the JASON study commented that one major exercise conducted in 2002—designed in part to test the technical capabilities of a set of biosensors—seemed "...to be little more than a demonstration of currently funded programs."[78]

In light of these shortcomings, this study was conducted as a "first-cut" evaluation of a system-of-systems approach. Rather than relying on just one data stream for information about a possible attack, this study has devised an integrated system with ten separate data streams and uses them in a modified wargame scenario to allow participants—who act as advisors to policymakers—to evaluate the value of the various reporting systems. Using a statistical analysis technique known as the analysis of variance (ANOVA)[79], the results of the wargame allow us to recommend suggested combinations of data streams—a system-of-systems—that would provide the highest value data to policymakers.

The wargame is concerned only with the release of a biological agent (man-made or naturally occurring) and thus focuses on the use of the data to indicate an outbreak of disease. Thus, it assumes a system-of-systems operating as detect-to-treat; although, it attempts to provide knowledge of an incident at the earliest possible time.

A depiction of the system-of-systems is shown in figure 20. In this graphic the various data sources are collated and subjected to analysis using a variety of proprietary and public tools and results in several outputs that are provided to policymakers. The

---

[77] Dena M. Bravata, et al., "Evaluating Detection and Diagnostic Decision Support Systems for Bioterrorism Response," *Emerging Infectious Diseases*, January 2004, Vol. 10, No. 1, 100-108: 100.
[78] JASON, "Biodetection Architectures," The MITRE Corporation, McLean, VA, February 2003: 2. See <http://www.fas.org/irp/agency/dod/jason/biodet.pdf>, accessed December 2003.
[79] See <http://trochim human.cornell.edu/tutorial/rehberg/popper.htm>, accessed March 2004.

generic, and in some cases specific, data fusion requirements have been reviewed by several researchers and are discussed in the publications of Lober, et al. (2002)[80] and Tech, et al. (2002).[81]

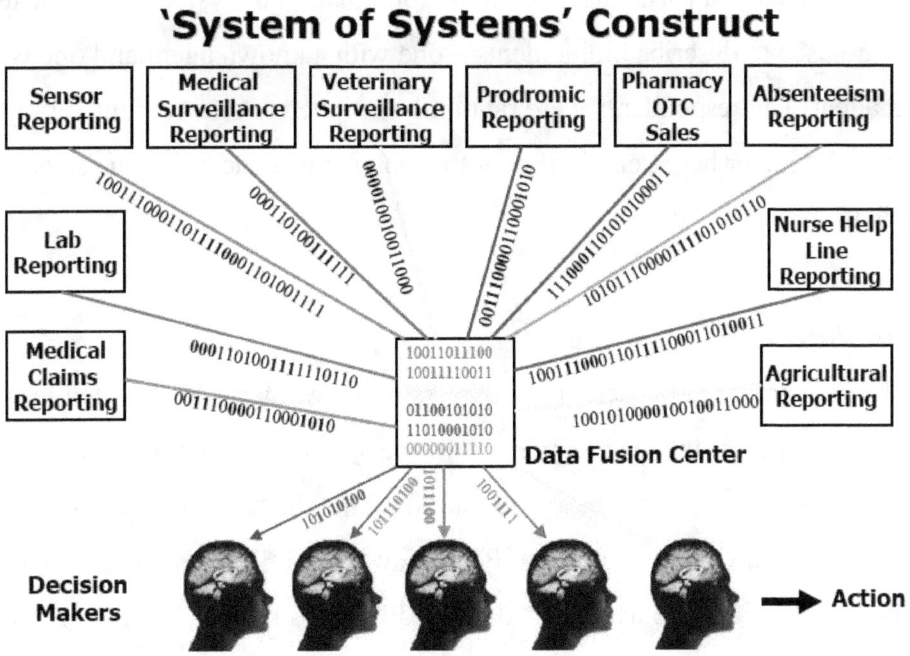

'System of Systems' Construct

Figure 20 – A system-of-systems approach to biomonitoring.

All of the proposed systems have been subjected to individual critical analysis and have generated claims about the specific performance characteristics that they provide. Against this background of claims about system performance in terms of sensitivity, specificity, range of agents detected, cost, utility, etc., there is a possibility (some would say good probability) that any selection process will discard a promising technology on the basis of limited value to decisionmaking. Moreover, a competitive analysis does not easily allow for the fact that the solution may best result from a combination of sensor systems or system-of-systems. Competitive "down-selection" is also subject to

---

[80] William B. Lober, et al., "Rountable on Bioterrorism Detection: Information System-based Surveillance," J Am Med Inform Assoc. 2002 Mar-Apr; 9(2): 105-15. See <http://www.pubmedcentral.nih.gov/articlerender.fcgi?tool=pubmed&pubmedid=11861622>, accessed January 2004.

[81] Tech et al., 2002.

considerable influence by the assessors – a system that reduces bias and that incorporates the rigor imposed by statistical analysis would appear to offer some advantages.

The wargame requires each respondent to use a pre-specified scoring system to assess the attributes of a particular system, or combination of systems. Two different scenarios are used to describe bioincidents—one with a known agent and one with an unknown agent. The respondents were randomly assigned to one of eight survey groups and asked about his or her strength of belief in statements concerning the sensor system under review.

The Wargame

The data provided to each of the players in the wargame is provided in Appendix A. For analysis, each of the ten components of the system-of-systems (see figure 20) was placed into one of three groups, based on their overall "theme." The Business Group consisted of Medical Claims Reporting, Pharmacy OTC Sales, and Absenteeism Reporting. The Medical Group consisted of Medical Surveillance Reporting, Veterinary Surveillance, Agricultural Reporting, and Nurse Help Line Calls. The Positive ID Group included Laboratory Reporting, Sensor Reporting, and Prodromic Reporting.

Each group and combination of groups was evaluated with respect to utility, trustworthiness, and resource requirements. (See Appendix B for full discussion of the experimental design and statistical analysis.)

Table 4 presents a simplified version of the ANOVA results (see Appendix B for the complete analysis). In table 4 the groups are shown in rank order, as judged by the wargame respondents. These are compared against a baseline. If a group—or combination of groups—is not listed, then it was not statistically significant from the baseline.

	Utility		*Trustworthiness*		*Resource Requirements*		*Overall*	
Scenario 1 Known Agent	Bus+Med+Pos.ID Bus+Pos.ID Med+Pos.ID Bus+Med Pos.ID Med		Pos.ID		Med+Pos.ID Pos.ID Med		Bus+Med+Pos.ID Bus+Pos.ID Med+Pos.ID Pos.ID Med	
Scenario 2 Unknown Agent	Med		Bus+Pos.ID		Med		Med	

Table 4 – Rank Ordering of Preferred Data Groups. "Bus" = Business Group; "Med" = Medical Group; "Pos.ID" = Positive Identification Group.

Veterinary Surveillance

Scenario 1 (Known Agent)

- Experts find the combination of all groups—Business+Medical+Positive ID—to have the most utility; more data are better than less.

- The Positive ID Group is viewed as the most trustworthy set of data, i.e., a sensor report that can confirm the identity of the agent is valuable.

- The combination of Medical+Positive ID Groups is the most demanding of resources; it will cost the most to operate.

- Overall, the combination of data from all three groups is preferred. Again, more data are better. Data derived solely from the Medical Group is better than the baseline, but least preferred.

Scenario 2 (Unknown Agent)

The unknown agent presents the more likely scenario. Whether naturally occurring or terrorist-induced, at least the early stages of a bioincident will include uncertainty as to the causative agent. In the event of an unknown agent attack, the experts conclude:

41

- Only Medical Group data will have any real utility.

- The combination of Business+Positive ID group reporting is seen as trustworthy. (This finding appears somewhat anomalous and will be further investigated in follow-on studies. It is likely that the manner in which the questions were asked in the study led to this finding.)

- Medical Group reporting was seen as requiring the most resources when dealing with an unknown agent. Given that Medical Group reporting is rated as having the most utility, it is likely that the most resources would be put into collecting the data.

- Overall, when faced with an unknown agent, experts prefer to have data from the Medical Group.

Known vs. Unknown Biological Agents

It was clear from the wargame that there was a considerable difference in the perceived value of surveillance systems when the biological agent is known rather than unknown and that these differences were statistically significant in the wargame. This is an expected result and is a consequence of the fact that most of the surveillance systems that are being developed and were featured in the wargame only detect or identify specific biological agents. Moreover, some systems have an even narrower spectrum of use in that they require the specific biological agents to conform to strict limits in terms of their behavior in the assay system.[82] If the particular agent does not conform, for example, due to changes in the cellular components expressed at the cell surface, then the system will not recognize the agent and will fail to provide a positive result. If, however, the agent under test is an unknown agent or the known agent has characteristics that are in some way anomalous and thus do not meet the system limits of specificity or even sensitivity, then the system may be deemed to be of limited value.

Even when the groups were combined, the results indicate that there is a considerable difference across systems when they are required to address anything other

---

[82] Margaret E. Kosal, "The Basics of Chemical and Biological Weapons Detectors," Center for Nonproliferation Studies, Monterey Institute of International Studies, Monterey, CA, November 24, 2003. See <http://cns miis.edu/pubs/week/031124 htm>, accessed March 2004.

than a known biological agent. Thus, it is not surprising that the wargame confirms that we have far to go in achieving a system that is resource-feasible, trustworthy, and useful for an unknown agent or a known agent that does not conform to the expected pattern.

Conclusions

In assessing these results, it is important to note that the size and scope of this evaluation provide sufficient data to suggest that further and more comprehensive analysis is warranted. For example, by grouping system components together into three categories, it is difficult to know if one reporting component is accounting for all of the response, or if it is evenly divided. (For example, is it just absenteeism reporting, or is it OTC and medical claims?) Nevertheless, this preliminary look provides good guidance for designing the next round of investigation. The next round also requires a larger respondent base, to help increase the usefulness of the statistics.

Those reservations aside, this study supports conclusions that the Federal Government should:

- Reassess efforts currently underway that attempt to capture data from absenteeism reporting, OTC pharmacy sales and medical claims reporting. Their value-added may not be worth the cost.

- Increase efforts to capture medical data. These efforts would include, but not be limited to, capturing data from doctors' offices and ER visits, as well as expanded veterinary and agricultural surveillance. In addition, increase data collection from medical website visits and nurse helplines. These data sources are valuable for early detection of both known and unknown agents.

- Reassess current plans to significantly increase the number of biosensors deployed as part of both the BioWatch[83] and Guardian[84] programs, in light of the limited value of sensors for detecting unknown agents. (See further discussion of this point immediately below.)

---

[83] "Government Provides Details of Bioterror Sensors in Cities," The Associated Press, Washington, DC, November 15, 2003.
[84] Gerry J. Gilmore, "'Guardian' Project to Bolster Force, Installation Security," American Forces Press Service, Washington, DC, May 8, 2003. See
<http://www.defenselink mil/news/May2003/n05082003_200305084 html>, accessed March 2004.

A Hot Idea

So what are we to make of all the statistical analysis of our wargame? For one, it is clear that, when dealing with a known agent, decisionmakers place considerable value on data that provide positive identification of pathogens. If fact, the Positive ID Group was the only statistically significant group that was viewed as trustworthy for a known agent. In the overall rating, they were the highest ranked individual group and were part of all the other system-of-system approaches that were statistically significant. This conclusion parallels the current focus on sensors as a component of our strategies for homeland security.

As a nation, we are investing a significant amount of money in sensors as part of the BioWatch program. Recent press reports have described a $60 million sensor network deployed in 31 cities with a proposed increase to $118 million for 2005 to cover additional cities.[85] Details of the program are understandably classified, but there has been local acknowledgment of sensors in New York, Washington, Chicago, Houston, San Francisco, San Diego and Boston.[86] Government officials will not confirm what agents the system screens for, but they say it is less than a dozen.[87] That most likely limits it to the candidate agents listed in CDC's Category A list.[88]

The BioWatch program has not been universally accepted and has plenty of critics. Calvin Chue, a researcher at Johns Hopkins University, points out that BioWatch would likely be effective only in detecting a major atmospheric release.[89] Small-scale attacks, or attacks delivered through food or water, that could result in hundreds—or

---

[85] Deborah Charles, "US Germ Detection System Active in 31 Cities," Reuters, Washington, DC, November 14, 2003. See <http://www.stevequayle.com/News.alert/03_Global/031117.bio.detection.html>, accessed December 2003.

[86] Sean Whaley, "Outdoor Sensors Take Time to Detect Dangers," *Las Vegas Review-Journal*, Carson City, NV, December 27, 2003. See <http://www.reviewjournal.com/lvrj_home/2003/Dec-27-Sat-2003/news/22881907.html>, accessed December 2003.

[87] Stephen Prior, PhD, National Security Health Policy Center at the Potomac Institute for Policy Studies, personal communication.

[88] The Center for Disease Control and Prevention has standardized a list of the most likely bioterrorism agents. These six agents, identified as "Category A" agents, are anthrax, botulism, plague, smallpox, tularemia, and viral hemorrhagic fevers.

[89] "Government Provides Details of Bioterror Sensors in Cities," The Associated Press, Washington, DC, November 15, 2003. See <http://www.nytimes.com/2003/11/16/national/16TERR.html?ex=1082520000&en=c69bb316d19aa9b7&ei=5070>, accessed December 2003.

thousands—of victims would probably go undetected. Moreover, as Chue comments, BioWatch sensors do not monitor any indoor environments.[90] In this case, an indoor release will only be detected once it leaves the building and encounters a BioWatch sensor. The Director of the Center for Biological Defense at the University of South Florida—Jacqueline Cattani—summed up her feelings as follows: "If you saw planes going over and releasing major clouds of this stuff, there's a chance that people would get suspicious a long time before anybody checked the filters."[91]

Other critics further emphasize the fact that the sensors only cover one-half of the U.S. population.[92] They point out problems with positioning of the sensors and question the amount of air sampled.[93] Additionally, they cite the high labor costs involved with the system.[94] The filters have to be collected and processed by trained laboratory technicians.

Even the proponents of the program make statements that are somewhat self-damning. One EPA executive was quoted as saying that *if* an attack were close enough to a sensor, authorities could know about it within 12 hours.[95] He rightly pointed out that 12 hours is quicker than if we waited for victims to develop symptoms.[96] However, that *if* is a major point and raises the issue of spatial distribution and placement, as well as meteorological conditions. It is worth noting that an EPA sensor was in place just blocks from the World Trade Center towers, but following the collapse on September 11th, it did not register the incident—only when the wind direction changed on September 12th did the sensor become "aware" of the incident.[97]

The problem of false alarms (positive or negative) must also be considered for the BioWatch sensor network. One senior government official was quoted as saying that of

---

[90] The Associated Press (Nov. 15, 2003).
[91] Ibid.
[92] David McGlinchey, "US Health Officials Highlight Surveillance Systems," *Global Security Newswire*, October 22, 2003. See <http://www nti.org/d_newswire/issues/2003/10/22/44363c15-eeed-40ec-8098-a9beae659465 html>, accessed December 2003.
[93] David W. Siegrist, *Technology-Based Biodefense*, Potomac Institute for Policy Studies, September 2003: 5. See <http://www.potomacinstitute.org/pubs/BTIII_Intro.pdf>, accessed March 2004.
[94] Charles (Nov. 2003).
[95] "Fed's Sniffing Devices Effective?" The Associated Press, Philadelphia, PA, July 17, 2003. See <http://www.cbsnews.com/stories/2003/07/17/tech/main563765.shtml>, accessed December 2003.
[96] Ibid.
[97] Stephen Prior, PhD, National Security Health Policy Center at the Potomac Institute for Policy Studies, personal communication.

the nearly 500 sensors nationwide, not one had ever raised a false alarm.[98] That is so statistically unlikely as to be considered impossible.

Even 500 sensors are far too few for the coverage sought by BioWatch. The Federal Government is seeking to increase the number of collectors per city. According to Congressional testimony by Department of Homeland Security Under Secretary for Science and Technology Dr. Charles McQueary in late February 2004, the "average" city covered by the current BioWatch program only has ten collectors.[99] Studies indicate that 40-60 would provide the optimal coverage. In addition, according to Dr. McQueary, cities have requested more collectors to cover key facilities, such as transit systems, airports, and stadiums.[100]

As noted above, the BioWatch Program is capable of screening for fewer than a dozen known agents (i.e. agents that are predesignated as potential bioterrorist weapons). What about dealing with unknown agents, or known agents that are genetically modified to beat the sensor systems? In our wargame analysis[101], one of the scenarios tested calls for an unknown agent. In this scenario, sensors only show up as statistically significant when in combination with business reporting—and then only when assessing trustworthiness. (And, as stated in the discussion of the analysis in the previous chapter, this is most likely an anomalous finding, based on how the questions were posed. Re-evaluation of this finding in the next iteration of this study is indicated.) In all other cases, medical reporting is the only statistically significant measure when dealing with an unknown agent. This is an anticipated result, as the Medical Group relied on symptom recognition and required no knowledge of the agent—in other words, unknown agents are identical to known agents.

What is the likelihood an unknown agent will be released on the population? Would it be viral? Bacterial? Contagious? Actually, the human population deals with this issue frequently. An analysis of pathogenic microbes and infectious diseases reveals that, on average, for the 30-year period between 1973 and 2003, one new (i.e., previously

---

[98] The Associated Press (Nov. 15, 2003).
[99] Charles E. McQueary, Statement for the Record, Before the U.S. House of Representatives Subcommittee on Cybersecurity, Science, and Research & Development, February 25, 2004: 6.
[100] Ibid., 20.
[101] See "Wargame," this document.

unknown) disease emerged annually.[102] These new diseases did not appear only in remote corners of the world. Some of the more memorable ones that impacted the continental United States and Canada include Legionnaires' disease (1977), HIV/AIDS (1981), West Nile virus (1999), and SARS (2003).

Returning to our wargame analysis, medical reporting also ranks high when dealing with known biological agents. This suggests that rather than spending additional money on a sensor-based system that has applicability primarily for known agents, why not expand the system to include a medical component that has applicability across both known and unknown agents? (While DHS is attempting to develop an integrated, real-time, human-animal-plant surveillance system, as part of its Bio-Surveillance Program Initiative, such a complete system is many years off.[103] The sensor-based BioWatch component is currently a major part of the initiative.) This medical approach seems particularly prudent, given that we are investing in a system designed to detect both the possible (known agents) and the probable (unknown agents). Such a system-of-systems would then be useful both against bioterrorism and for general public health.

The JASON study noted that we already have roughly 300 million biosensors in the U.S.—our population.[104] Not only do humans sample the air when we breathe, but we also concentrate the sample, and our immune systems and innate responses to insult from biological challenges act as a detector. These responses are both highly specific and highly sensitive. If a sample of the population were monitored for the first signs of symptoms of challenge (regardless of route of infection), then the information gleaned could also be timely.

On the basis of these observations and our wargame analysis of the utility of the existing technology solutions, we propose that policemen, firemen, and mail carriers be used as a sentinel population to monitor for possible outbreaks of known or unknown agents. Statistically, they provide a better sample of air than stationary collectors, because they are more uniformly distributed across a metropolitan area. Unlike the current

---

[102] "US ignoring dangerous diseases during war on terror, says researcher," Canadian Press. See <http://story news.yahoo.com/news?tmpl=story&u=/cpress/20040429/ca_pr_on_he/health_bio_terror_conf erence&cid=2155&ncid=2155>, accessed May 2004.

[103] Tommy Thompson, Bill Frist, Press Conference, January 29, 2004. See <http://www.dhs.gov/dhspublic/display?content=3093>, accessed March 2004.

[104] JASON, "Biodetection Architectures," The MITRE Corporation, McLean, VA, February 2003: 14. See <http://www.fas.org/irp/agency/dod/jason/biodet.pdf>, accessed December 2003.

systems, they are not subject to the vagaries of microclimates. Also, during the course of their duties they encounter both outdoor and indoor environments. Their daily routes and movements are fairly well-defined, making it easy to pinpoint affected areas.

Although the performance characteristics of the BioWatch collectors remain classified, it is relatively easy to estimate their capacity to sample the air. Current EPA "high-volume" environmental air samplers are rated at taking in 40-60 cubic feet of air per minute (cfm).[105] Assume a "very-high-volume" capacity for the BioWatch collectors at 100 cfm (equal to 2,832 liters/minute). At that rate, the collectors sample the equivalent of one large room (24.5'x24.5'x10') per hour, or approximately 170,000 liters of air.

Take the metropolitan Washington, DC workforce as a sentinel population. There are 12,110 police and sheriff's patrol officers, 4,900 firefighters, 400 EMS workers, 5,940 mail carriers, and 150 parking enforcement workers for a total population of 23,500 people.[106] Assume an average breathing rate of .5 liters per breath[107], with 16 breaths per minute.[108] (That is a conservative breathing rate, as it is derived from "at rest" figures. Given their level of physical activity, this population is very likely breathing more than that.) So, in one hour each member of this sentinel population samples 480 liters of air. Per hour, the entire workforce samples 11,280,000 liters of air. That is roughly the equivalent of 66 very-high-volume samplers.

---

[105] United States Environmental Protection Agency, "Particulate Emission Measurements from Controlled Construction Activities," Prepared by National Risk Management Research Laboratory, Research Triangle Park, NC, April 2001: 14. See <http://www.epa.gov/appcdwww/apb/R-01-031/EPA-600-R-01-031.pdf>, accessed January 2004.

[106] 2002 Metropolitan Area Occupational Employment and Wage Estimates, Washington, DC-MD-VA-WV PMSA. See <http://www.bls.gov/oes/2002/oes_8840.htm#b43-0000>, accessed February 2004.

[107] See <http://www.nqinc.com/faq16.html>, accessed March 2004.

[108] See <http://transitiontoparenthood.com/ttp/parented/pregnancy/abdobreathe.htm>, accessed March 2004.

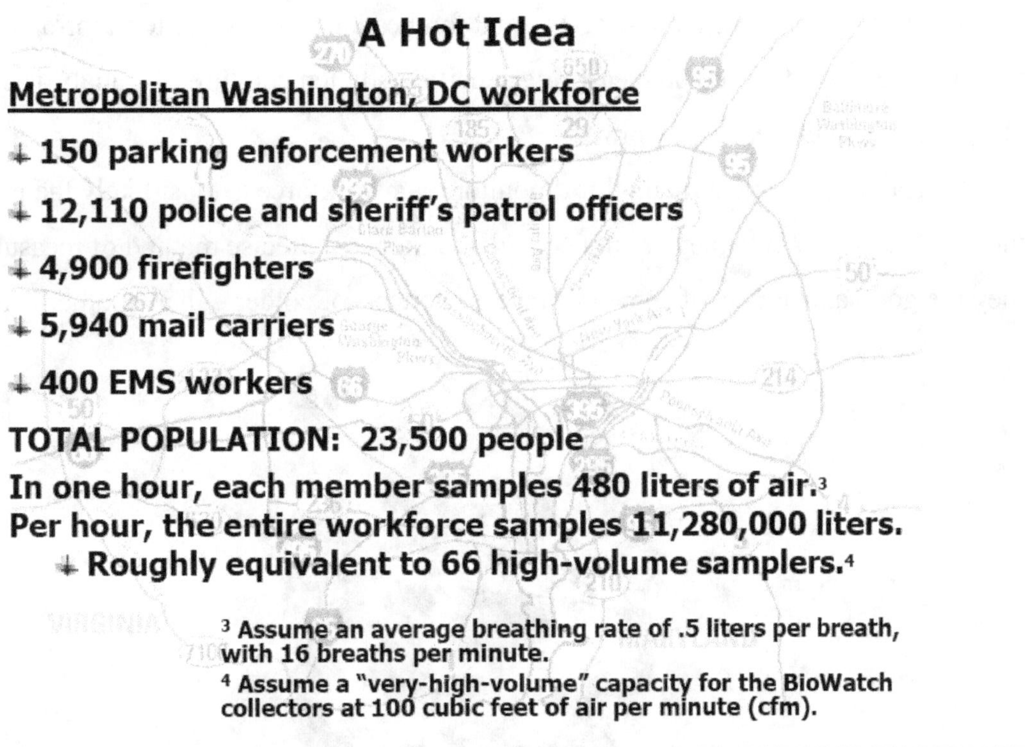

A Hot Idea

Metropolitan Washington, DC workforce

+ **150 parking enforcement workers**

+ **12,110 police and sheriff's patrol officers**

+ **4,900 firefighters**

+ **5,940 mail carriers**

+ **400 EMS workers**

TOTAL POPULATION: 23,500 people

In one hour, each member samples 480 liters of air.[3]

Per hour, the entire workforce samples 11,280,000 liters.

+ **Roughly equivalent to 66 high-volume samplers.[4]**

[3] Assume an average breathing rate of .5 liters per breath, with 16 breaths per minute.

[4] Assume a "very-high-volume" capacity for the BioWatch collectors at 100 cubic feet of air per minute (cfm).

Figure 21 – The selected workforce is uniformly distributed across the city and encounters both indoor and outdoor environments.

These samplers are highly specific and highly sensitive, and they have no false positives. What data could be efficiently and economically collected from them, to assess the results of their sampling? Considerable work is being done on identifying various blood components that would indicate the earliest signs of disease. These components would be non-specific and alert us that the body is in the early stages of an immune response. Further blood tests would be needed to identify the agent. Urine, sweat, or breath might also be tested. The perfection of these early identification tests (often referred to as prodromal states) is still well in the future. Plus, there will likely be considerable logistics and economic factors to consider.

At the moment, of all the potential parameters that can be readily measured, body temperature offers the most efficient and economical approach. Moreover, the wargame results indicated medical information is the preferred data when dealing with an unknown agent. Body temperature is a primary piece of medical information.

The technology for measuring body temperature was widely used during the recent SARS epidemic. Various approaches were used, including infrared thermal imagers, oral or ear fever thermometers, and forehead, or temporal artery, infrared thermometers. All focused on measuring specific body temperatures for analysis of any elevation above the expected norm.

The most efficient method for monitoring a workforce is most likely the infrared thermal imagers. While they are not considered the most precise method of measurement, they are good at detecting if someone appears hotter—or colder—than another person.

Figure 22 – In general, the immune system will register the presence of a foreign agent by increasing the body temperature within a matter of hours. These increases could be easily detected by infrared thermal imagers, such as those shown here being used during the recent SARS outbreak.

Every member of a shift could be scanned as he/she came to work. They could all be scanned again at the end of their shift. If members of a shift were exposed near the end of their tour, their temperature might not be elevated by the time they are scanned prior to going home. However, it is likely that the next shift coming on will also be exposed—this time at the beginning of their shift—and will have a measurable temperature increase by the end of their shift. (Baseline data could be stored on all employees and individual temperature variations could be easily accounted for.)

In general, the immune system will register the presence of a foreign agent by increasing the body temperature within a matter of hours.[109] One person showing an elevated temperature might not warrant further investigation, but a cluster would suggest the need for further blood tests.

Given the cost of purchasing and maintaining the systems proposed under BioWatch, it seems worth the time and effort to conduct a test of the feasibility of thermal imaging as a medical monitoring technique. It certainly appears to offer a rapid, simple, and cost effective capability for disease detection that is currently ignored.

By contrast, the military presents a different set of problems, when considering the detection of bioagents. Forces in the field need a standoff detection capability. Not only does it provide them with a detect-to-warn capacity that would not be practical in a civilian setting, it also provides them with a reconnaissance capability. There is no argument that the continued development and deployment of biosensors is critical for combat operations.

The Department of Defense is scheduled to award a $1.1 billion contract for its Installation Protection Plan (IPP).[110] Ultimately, the IPP is to establish a network of chemical, biological, radiological and nuclear-detection sensors at 200 military installations worldwide.[111] While the chemical, radiological and nuclear-detection sensors may be important and necessary parts of the Guardian program, this study suggests that the biosensor component might be modified to include thermal imaging. As with the BioWatch program, a short test of this idea seems fiscally responsible, prior to embarking on this five-year effort.

---

[109] Christopher Green, MD, PhD, Wayne State University Medical School, personal communication.
[110] "Lockheed Martin, CACI, EAI, Fluor and IEM Team For DoD Installation Protection Program," PRNewswire, CACI International, Inc., Manassas, VA, February 20, 2004. See <http://www.stockhouse.ca/news/news.asp?newsid=2161668&tick=>, accessed March 2004.
[111] Ibid.

Appendix A – The Wargame

Background

In this wargame we are asking you to play the role of an advisor to the decisionmakers that are charged with responding to the bioincident described in the two scenarios. For those of you who occupy positions in which providing advice to decisionmakers is part of your day-to-day activities, please play the war game as yourself. For those who do not routinely interact with decisionmakers, please play the war game as if you have been contacted for your advice based on the particular skill set that you have developed and the work that you currently perform. The key in this wargame is to gather a large number of responses; the analysis will enable us to draw conclusions from the wargame inputs.

Game Play

You will read two brief scenarios, each with a specific array of reporting mechanisms that you would be able to task. Use the Excel spreadsheet to evaluate how useful the set of sensors/reporting mechanisms is to reacting to the incident. For the purposes of this evaluation, we also provide characteristics of the biological agent that would be unknown at the time of the incident. You can use these agent characteristics to help score the sensor set; e.g., a set would score fairly well if it can quickly, reliably and cheaply provide the agent's traits. *Remember that you are scoring the sensors in their entirety, i.e. as a "package deal."*

The essence of the wargame is for you to assess the bioincident and the sensor systems that are described and for you to determine how well the sensor set that is available to you, in the scenario under study, would support you advising a decision maker to take action. At this stage it is not important to discriminate between action that would be considered as "investigative" or as a full "response" – this will be evaluated in future iterations of the wargame.

Scenarios

We are asking that you consider two scenarios. In the first scenario the biological agent conforms to our current threat list and results in infections in humans that correspondingly conform to expected symptoms. Examples in this case would include anthrax, botulinum toxin, plague, etc. In the second scenario we are asking you to consider a biological agent that does not form part of our current threat list or behaves in an abnormal manner in humans. Examples here would include a bio-engineered strain of anthrax or plague, or an agent not on our current priority list but that could still be used in attacks on humans (examples might include Q-fever).

The reason for the inclusion of the second scenario is that some of the sensor systems that we are asking you to evaluate are limited in terms of detection capacity to a very small number of agents – specifically those that are a priority on the threat list. Their utility may be similarly limited. We are keeping most of the scenario parameters constant and have provided a simple outline to limit the amount of reading on the part of the respondents – this will mean that for some people the scenarios are somewhat unclear. In this case the scenarios will likely correspond to real life – in any bioincident we will have access to much less information than we would like to have before being asked to provide an opinion. In this wargame please use your best judgment based on the data provided.

In both scenarios the initial release occurred in a major metropolitan area of the United States. The incident occurred approximately four days prior to the time of the war game – i.e., the biological agent will have had time to incubate and it could be anticipated that those who have been exposed (and possibly their immediate contacts) will be exhibiting symptoms. The sensor systems or reporting systems will have been triggered and data will have been reported to various authorities. Specific information will still be unavailable or not yet confirmed by a federal agency.

Scenario #1 – "Known Biological Agent"

In this scenario the biological agent affects humans and has the following characteristics:
- Known agent – i.e. on threat lists and not genetically-modified.
- Infectious or contagious – mode of transmission not yet identified.
- Transmission rate (humans) ~60%.
- Symptoms/diagnosis conform to threat list data.

- Specific sensor technologies or tests will produce positive result.

Scenario #2 – "Unknown Biological Agent"

In this scenario the biological agent affects humans and has the following characteristics:
- Unknown agent – i.e. not on threat lists – newly emergent or modified agent.
- Infectious or contagious – mode of transmission not yet identified.
- Transmission rate (humans) ~60%.
- Symptoms do not conform to any known agent.
- Specific sensor technologies or tests will produce negative or ambiguous results.

Sensor Groups Under Review

In order to make the first iteration of the wargame statistically manageable and not too onerous for the respondents, we have grouped the sensor systems to create three sets of technologies that will be assessed for their contribution to a detection capability. For each of the groups we have included sensors that are currently considered candidates as detector systems following a biological release. The groups each have sensor systems with a range of characteristics with respect to sensitivity, specificity, cost, etc. In each group we have included at least one sensor system that is capable (when fully deployed) of providing data very quickly after release of the agent and at least one system that will provide more specific data on the characteristics of the agent (but not necessarily the exact identity) some time after the incident is established. The sensor groups are shown below:

Group A
Absenteeism rates from selected (normalized) populations.
OTC pharmacy sales (electronic data – commercial sites).
Medical claims reporting.

Group B
Doctors Office & ER reporting (electronic data – e.g. ICD-9).
Veterinary & agricultural reporting from sentinel populations.
Calls/web-enquiries to Nurse Helplines

Group C
Sensor reports from fixed sites.
Reports of prodromic symptoms for selected disease states.
Laboratory test results.

The wargame is designed to assess the value of the sensor groups to the respondent (not anyone else) when that person is functioning as a decision maker or as an advisor to a decision maker. In this case decision maker refers to a person's ability to take "trigger" action or actions that will impact the emerging incident. We are asking that you provide your input on assessment of the sensor group characteristics for the groups that have been randomly assigned to you – the statistical plan will enable us to gather data on all of the sensor groups and combinations of sensor groups that will drive the outcome for the wargame.

Sensor Group Attributes and Scoring

We have selected a range of attributes that will be assessed for each of the sensor groups by our wargame. They include all of the key characteristics and will be assessed based on their use as "Attribution Statements" to which the respondents are asked to provide responses based on their "strength of belief" in the statements when they are applied to the sensor groups. The attributes that are being assessed are shown below:

Utility
- Signal availability
- Timeliness
- Actionable

Trustworthiness
- Sensitivity
- Specificity
- Reliability

Resource requirements
- Cost
- Sustainability
- Mode of collection
- Data source

Wargame Example Score Sheet (Response Group 1)

System-of-Systems 'wargame'
Score Sheet – Respondent Group Number 1

Notes:
Please read the attached notes before completing this survey.
You are being asked to complete the survey on the basis of random assignment to one of eight groups.
Your input is limited to the 'Survey Group' (or Groups) shown below

Scoring In each case the 'scoring' will be based on the following scale:

Survey Score	Response to Statement
1	Strongly disagree
2	Disagree
3	Neither agree or disagree
4	Agree
5	Strongly agree

Your Responses:

Respondent Group Number	1
Scenario #1 - Known Agent	Sensor Group C only
Scenario #2 - Unknown Agent	Sensor Groups A+B+C

Score-Sheet

Respondent Group Number 1

Statement	Scenario #1 Sensor Group C Score	Scenario #2 Sensor Group A+B+C Score
The sensor signals are available at early stages of agent release or human exposure		
The set of information is timely and provides advanced warning		
The set of information is actionable rather than informational		
The system is highly sensitive to the agent (high recall)		
The system will have high specificity (low percentage of false negatives)		
The system output and results are reliable and subject to validation		
The benefits of the system outweigh its cost		
The system requires little or no input to provide data 24/7/365		
The technology requirements are easily met using current technology		
The data requirements of the system are easily met through existing data sets		

Appendix B – Methodology & Statistical Analysis

Survey Design

We identified ten desirable characteristics of a warning system for biological agents and grouped these into three categories [Table B.1]. Survey respondents evaluated Warning Systems on these characteristics using a 1-to-5 Likert scale of agreement. Respondents were solicited via email and the scoring was conducted in a Microsoft Excel spreadsheet attachment.

Experimental Design

For each of the two scenarios, Known Agent and Unknown Agent, an experimental design was conducted. First, the authors drafted a list of nine potential attack indicators placed into three groups [table B.2] and a list of expert evaluators with various backgrounds. Rather than have experts score each Indicator Group one at a time (as in a typical survey), we implemented a full factorial design with the Indicator Groups as factors. That is, experts were randomly assigned to assess one of eight Systems, varying from a baseline of no indicators to a full system of all three Groups: Business, Medical, and Sensor indicators. This design allows us to assess the value of Group combinations, i.e. do Indicator Groups complement or duplicate each other's information for first responders?

Data Analysis

The authors sent surveys to 132 experts and further instructed them to distribute the instrument to other experts. This yielded 49 respondents, each of which analyzed one Warning System for each Scenario. Composite scores for Utility, Trustworthiness, and Resource Requirements were created by averaging the characteristics for each. Averaging the three composite scores created an Overall score. Table B.3 shows the raw scores and a measure of the components' consistency/reliability, Cronbach's Alpha. Note that the reliability for Trustworthiness is marginal (.65), suggesting that measurement of this element needs refinement in future studies.

Analysis of Variance (ANOVA) was performed on the four composite scores for each scenario. This statistical analysis can detect patterns in the eight combinations of Indicator Groups to isolate those that drive better scores. For example, examining the Utility Scores for the Unknown Agent in table B.3, we see that Systems that include Medical information score higher than those without Medical data do. These results are reflected in the ANOVA coefficients [table B.4] and the Systems that the models indicate as better than the baseline [table B.5].

Notes

Fit statistics and coefficients for the Known Agent models are generally larger than those of the Unknown Agent models. This suggests that warning for Unknown Agents is more difficult [table B.4].

Additional warning indicators can be added to the Preferred Systems in table 5, but doing so will not provide responders & policymakers with statistically-significant value-added.

This study was not designed to isolate the individual effects of the components of the three Indicator Groups (Medical, Business and Positive ID). This will require additional study-- for example, a larger sample employing a fractional factorial design with individual indicators as factors.

Table B.1 – Desirable Characteristics of Warning Systems & Category Assignments

Characteristics of Warning System and its Output	Category
The sensor signals are available at early stages of agent release or human exposure.	Utility
The set of information is timely and provides advanced warning.	Utility
The set of information is actionable rather than informational.	Utility
The system output and results are reliable and subject to validation.	Trustworthiness
The system is highly sensitive to the agent (high recall).	Trustworthiness
The system will have high specificity (low percentage of false negatives).	Trustworthiness
The benefits of the system outweigh its cost.	Resource Requirements
The system requires little or no input to provide an alarm function 24/7/365.	Resource Requirements
The technology requirements of the system are easily met using current technology.	Resource Requirements
The data requirements of the system are easily met through existing data sets.	Resource Requirements

Table B.2 – Biological Attack Indicators & Group Assignments

Biological Attack Indicators	Group
Medical Claims Reporting	Business
Pharmacy OTC Sales	Business
Absenteeism Reporting	Business
Medical Surveillance: Doctors' Offices	Medical
Medical Surveillance: ER	Medical
Veterinary and Agricultural Surveillance	Medical
Nurse Help Line Calls	Medical
Laboratory Reporting	Positive ID
Sensor Reporting	Positive ID
Prodromic Reporting	Positive ID

Table B.3 – Means, Sample Sizes & Reliability Scores for Composite Scores

	Indicator Groups	N	COMPOSITE SCORES			
			Utility (α=.84)	Trust (α=.65)	Resource Requirements (α=.76)	Overall (α=.86)
	Baseline (no indicators)	11	1.8	2.2	2.4	2.1
Scenario 1 Known Agent						
	Business	5	1.6	2.1	2.8	2.2
	Medical	7	2.8	2.5	3.7	3.0
	Positive ID	5	3.3	3.6	3.2	3.4
	Business + Medical	5	1.9	1.6	2.9	2.1
	Business + Pos. ID	6	3.6	3.5	4.0	3.7
	Medical + Pos. ID	9	3.4	3.4	3.6	3.5
	Business + Medical + Positive ID	6	4.6	4.2	3.9	4.2
Scenario 2 Unknown Agent						
	Business	6	1.8	1.9	2.7	2.1
	Medical	6	2.6	2.4	3.0	2.7
	Positive ID	6	2.2	2.1	2.2	2.1
	Business + Medical	7	2.5	2.2	3.5	2.7
	Business + Pos. ID	5	2.2	3.0	3.0	2.7
	Medical + Pos. ID	5	2.6	2.3	3.1	2.7
	Business + Medical + Positive ID	5	3.5	3.5	3.0	3.4

Table B.4 – Simplified ANOVA Models: Coefficients and R^2

	Utility		Trustworthiness		Resource Requirements		Overall	
Scenario 1 Known Agent								
	R^2	.58	R^2	.52	R^2	.25	R^2	.56
	Intercept	1.9**	Intercept	2.1**	Intercept	2.7*	Intercept	2.3**
	Sensor	1.0**	Sensor	1.5**	Sensor	0.7*	Sensor	0.9**
	Med	0.6**			Med	0.5*	Med	0.4*
	Bus	-0.5					Bus	-0.4
	Bus*Sensor	1.3**					Bus*Sensor	0.9*
Scenario 2 Unknown Agent								
	R^2	.14	R^2	.26	R^2	.13	R^2	.13
	Intercept	2.0**	Intercept	2.2**	Intercept	2.5**	Intercept	2.2**
	Med	0.8**	Sensor	-0.0	Med	0.7**	Med	0.6**
			Bus	-0.2				
			Bus*Sensor	1.3**				

Statistical Significance: * <.05; **<.01
Non-significant factors were removed from models unless higher-order interactions were significant.

Table B.5 – Systems Preferred over Baseline and Predicted Scores

	Utility		Trustworthiness		Resource Requirements		Overall	
Scenario 1 Known Agent	Bus+Med+Sensor	4.4	Sensor	3.7	Med+Sensor	3.9	Bus+Med+Sensor	4.2
	Bus+Sensor	3.8			Sensor	3.4	Bus+Sensor	3.7
	Med+Sensor	3.6			Med	3.1	Med+Sensor	3.6
	Bus+Med	2.1					Sensor	3.2
	Sensor	2.9					Med	2.7
	Med	2.6						
Scenario 2 Unknown Agent	Med	2.7	Bus+Sensor	3.3	Med	3.2	Med	2.8

Note: If an Indicator Group is not present in the listed System, there is no statistically significant value-added.